Accelerated Lattice Boltzmann Model for Colloidal Suspensions

Hassan Farhat • Joon Sang Lee
Sasidhar Kondaraju

Accelerated Lattice Boltzmann Model for Colloidal Suspensions

Rheology and Interface Morphology

Springer

Hassan Farhat
Department of Mechanical Engineering
Wayne State University
Detroit, MI, USA

Joon Sang Lee
Yonsei University
Seoul, Korea, Republic of (South Korea)

Sasidhar Kondaraju
Department of Mechanical Engineering
Indian Institute of Technology, Delhi
New Delhi, India

Videos to this book can be accessed at http://www.springerimages.com/videos/978-1-4614-7223-0

ISBN 978-1-4899-7401-3 ISBN 978-1-4899-7402-0 (eBook)
DOI 10.1007/978-1-4899-7402-0
Springer New York Heidelberg Dordrecht London

Library of Congress Control Number: 2014933943

Printed on acid-free paper

Springer is part of Springer Science+Business Media (www.springer.com)

Preface

The study of immiscible droplet-based systems, colloids, and interfaces at microscale levels is not only intriguing but also difficult to understand using direct experimental measurements. The inception of the lattice Boltzmann method (LBM) has made the study of liquid–liquid suspensions more attractive due to its simplicity in handling the droplets interfaces under a variety of flow conditions. Although books have been previously written to explain the various LBM schemes and their applications, less attention was given to the modeling of multicomponent systems. Inherent interfacial formation in LBM along with the new improvements made to the model, such as moving multiblock grid, surfactant covered droplets, and suppression of coalescence between neighboring droplets, made it possible to utilize the model for the study of colloids in applications related to rheology and blood flow under certain conditions.

The objective of this book is to provide an efficient LBM-based CFD model, capable of solving complex problems related in general to liquid–liquid based macrocolloids, and in particular to biological suspensions under specific conditions. Throughout the book, we have provided the pseudocodes which cover the new improvements made to existing multicomponent LBM routines.

In Chap. 3 the concept of the migrating multiblock for accelerating the LBM solutions is explained. This covers the single component, the Gunstensen multicomponent, and the particle-interaction-potential LBM. The migrating multiblock concept is tested and validated in 2D and 3D geometries for flow conditions such as vortex shedding, shear lift of a neutrally buoyant droplet, buoyancy of bubbles, and settling droplets on a horizontal wall.

Chapter 4 presents a Gunstensen-based hybrid LBM-finite difference model for the study of surfactant-covered droplets. The coupling of the model is realized through the LBM velocity field and the surfactant equation of state. The model is tested and validated by studying the effects of surfactants on the flow deformation of a droplet in simple shear flow, uniaxial extensional flow, and under buoyancy.

Chapter 5 introduces a novel method for the suppression of coalescence in the 2D and 3D Gunstensen LBM and shows some test cases of the model. The importance of the suppression of coalescence module is in providing a qualitative representation of the interparticle forces which act on the interfaces of the approaching droplets. This allows the study of the rheology of colloids which results were validated by a comparison with some empirical solutions.

Chapter 6 discusses the RBC deformation in the microvasculature. A heuristic approach for simulating the RBC as surfactant-covered droplet is presented to assess the validity of the concept. The velocity and deformation of the droplet are studied as functions of the interfacial tension. The Fahraeus and the Fahraeus–Lindqvist effects are simulated and analyzed.

Detroit, MI, USA Hassan Farhat
Seoul, South Korea Joon Sang Lee
New Delhi, India Sasidhar Kondaraju

Contents

Chapter 1
Introduction

Abstract This chapter introduces briefly colloids, and in particular emulsions and their importance in the food, medical, cosmetic, polymer, water purification, and pharmaceutical industries. It also highlights the relation between biological suspensions and colloids, which can be studied by using the same approach. The chapter identifies several numerical method used traditionally for studying colloids and biological suspensions.

1.1 Colloids

Colloids are chemical mixtures with one substance dispersed into the other. Colloids are characterized by a high surface to volume ratio or surface to mass ratio. This enables their interparticle forces to alter the gravity effects. The focus here is on emulsions, one type of colloids in which the continuous medium and the dispersed phase are both immiscible liquids, although the presented model can handle with comfort other types of colloids such as foams and liquid aerosols.

The study of droplet-based immiscible mixtures is of high interest for broad range of research works in the food, medical, cosmetic, polymer, water purification, and pharmaceutical industries. Macroemulsions such as water-in-oil (W/O) or oil-in-water (O/W) are indispensable in the makeup of a great number of frequently used products. To satisfy a large and diverse market demands, polymer manufacturing industries strive constantly to supply new blends with enhanced thermal and mechanical behavior. Droplet-based microfluidic systems provide a highly controllable platform for applications, such as micro-reactors, drug delivery systems, and information carriers on microfluidic chips containing digital logic gates.

The control of the droplet size during the production of such systems is of extreme importance. Conventional methods such as rotor-stator or more current methods like membrane emulsification (van der Graaf 2006) are used to control the

H. Farhat et al., *Accelerated Lattice Boltzmann Model for Colloidal Suspensions: Rheology and Interface Morphology*, DOI 10.1007/978-1-4899-7402-0_1,

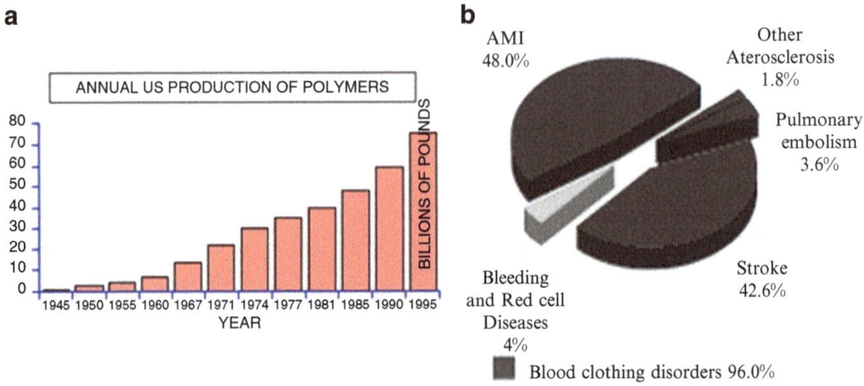

Fig. 1.1 (**a**) Annual US polymers production showing an exponential growth. (**b**) Death percentage rate in the USA in 2004 due to blood-related diseases

droplet size of an emulsion during formation. The required morphology of polymer blends is dependent on the rapid establishment of the equilibrium between droplet break up and coalescence (Sundararaj and Macosko 1995). Microfluidic droplets have to be generated in a specific frequency and sizes in order to deliver stable and repetitive functions.

The control of the droplet size is accomplished mainly by the addition of surface acting agents (surfactants), which adhere to the droplet interface and reduce the interfacial tension. Surfactants play also an important role in suppressing coalescence of the dispersed phase, and they affect the rheology of the droplet-based immiscible mixtures as a result of the intricate interplay of the evolution of surfactant distribution, drop deformation, and the bulk flow (Vlahovska et al. 2005).

Biological fluids such as saliva, urine, and blood are macrocolloids by nature. One of the most interesting biological fluids is blood. Blood is a biological suspension composed of 55 % plasma and 45 % formed elements of which 99.5 % are red blood cells (RBC), 0.13 % white blood cells (WBC), and 4.9 % platelets. In the capillaries the membrane of the RBC fluidizes under pressure, making it feasible to approximate the cell as a surfactant-covered droplet.

Historically polymer production in the USA was characterized by an exponential growth as shown in Fig. 1.1a. 40 % of this production serves as functional materials such as cosmetics, pharmaceuticals, printing ink, paints, and super-absorbers in hygienic products.

Another area of relevance is the investigation of biological suspensions, in particular blood and blood related diseases. In 2004 an estimated 223,000 deaths were due to blood diseases. 214,000 were caused by blood-clotting disorders a major contributor to acute myocardial infarction (AMI) and cerebrovascular diseases. Nine thousand were attributed to RBC and bleeding disorders. Blood clotting disorders were expected to cost the nation's economy an amount of 105 billion, in addition to 14 billion dollars due to other blood diseases in 2008 (www.nhlbi.nih.gov/about/factbook-07). Figure 1.1b shows the mortality rate due to blood-related diseases in 2004.

1.2 Numerical Methods for Colloidal Studies

The last couple of decades have witnessed a considerable advancement in the computer technology which manifested itself by an exponential growth in computing powers. This made it possible to explore the full potential of an already matured branch of mathematics (numerical methods) which became a primary tool for the study of a variety of fluid problems. Colloids and biological suspensions had been the subject of investigations by a great number of numerical researchers who used large diversity of methods such as the boundary integral (Milliken et al. 1993; Li and Pozridikis 1997; Eggleton et al. 2001; Feigl et al. 2007), the volume of fluid method (Drumright-Clarke 2002; Drumright-Clarke and Renardy 2004), the finite element method (Kruijt-Stegeman et al. 2004), the immersed boundary method (Lai et al. 2008), and the lattice Boltzmann method (LBM) (van der sman and van der Graaf 2006). These methods were used for the study of colloids. Other methods such as the immersed finite element method (Liu and Liu 2006), the particle method (Tsubota et al. 2006), the LBM (Dupin et al. 2003; Dupin et al. 2005; Sun and Munn 2005), and the hybrid LBM (Dupin et al. 2007) were used for the study of blood flows. In this book an accelerated multicomponent LBM scheme with incorporated surfactants effects will be presented and used for the study of emulsions and biological fluids.

Chapter 2
Overview of Related Studies

Abstract This chapter reviews the lattice Boltzmann method (LBM) for single phase and multiphase flows with a focus on the Gunstensen and the Shan and Chen models. It also explains the LBM multiblock scheme, which allows the iteration to occur simultaneously in a simulation domain consisting of smooth and coarse grids. This chapter also covers surfactant-laden droplets behavior from both experimental and numerical point of view, and it highlights the main physical characteristics of this phenomenon. Emulsions rheology is reviewed here, where several empirical formulations for the calculation of the relative viscosity of non-Newtonian fluids are highlighted. Explanation for the direct calculation of emulsions' viscosity in numerical simulations is also presented. Red blood cells as major constituent of blood determines its macrocolloid like behavior. This behavior is mainly influenced by the RBC properties, which are summarized at the end of this chapter.

2.1 The Lattice Boltzmann Method

Computational fluid dynamics (CFD) provides a powerful tool for the study of complex multiphase and multicomponent flows. Drop formation, deformation, coalescence, and brake-up continue to be the focus of many research works devoted for a better understanding of microfluidic, colloids and polymers properties. Among many CFD tools, the lattice Boltzmann method (LBM) has attracted some attention during the last couple of decades due to the simplicity of its algorithm, stability, and parallelism.

H. Farhat et al., *Accelerated Lattice Boltzmann Model for Colloidal Suspensions: Rheology and Interface Morphology*, DOI 10.1007/978-1-4899-7402-0_2,

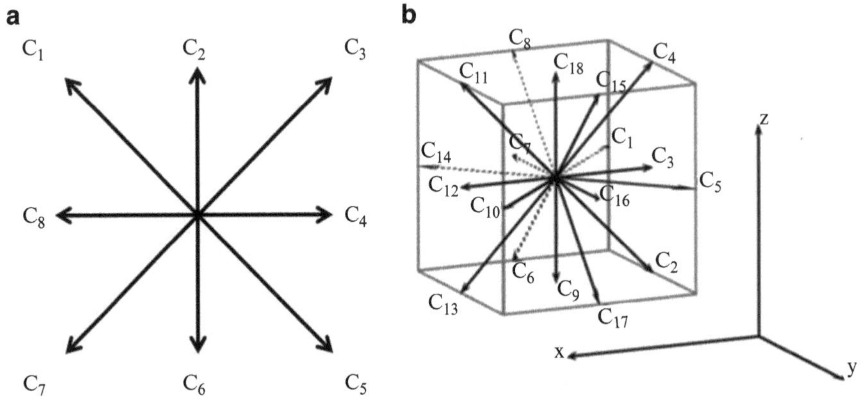

Fig. 2.1 (**a**) Velocity vectors for the D2Q9 and (**b**) for the D3Q19 lattice Boltzmann method used in this study

2.1.1 *The Single Component LBM*

The Bhatnagar-Gross-Krook (BGK) LBM is an alternative computational technique used for solving a broad range of fluid problems. The isothermal, single-relaxation model is derived from the following Boltzmann kinetic equation (Yu et al. 2002):

$$\frac{df}{dt} + \boldsymbol{\xi} \cdot \nabla f = -\frac{1}{\lambda}(f - f^{eq}) \tag{2.1}$$

where f is the density distribution function, ξ is the macroscopic velocity, f^{eq} is the equilibrium distribution function, and λ is the physical relaxation time. Equation (2.1) is first discretized by using a set of velocities ξ_i confined to a finite number of directions and this leads to the following equation:

$$\frac{df_i}{dt} + \boldsymbol{\xi}_i \cdot \nabla f_i = -\frac{1}{\lambda}(f_i - f_i^{eq}) \tag{2.2}$$

The LBM is based on a set of equivalent Cartesian velocities. The D2Q9 BGK described here has nine velocity direction vectors (lattice links) shown in Fig. 2.1a with the following endpoint coordinates:

$$\begin{gathered} e_0(0,0);e_1(-1,1);e_2(0,1);e_3(1,1);e_4(1,0)e_5(1,-1);e_6(0,-1); \\ e_7(-1,-1);e_8(-1.0) \end{gathered} \tag{2.3}$$

Figure 2.1b shows the lattice links for the D3Q19 model. Equation (2.2) is further discretized in the lattice space and time and this leads to the following:

$$f_i(\mathbf{x} + \mathbf{c}_i\delta_t, t + \delta_t) - f_i(\mathbf{x}, t) = -\frac{1}{\tau}\left[f_i(\mathbf{x}, t) - f_i^{eq}(\mathbf{x})\right] \tag{2.4}$$

The lattice space δ_x and the lattice time step δ_t are taken as unity and their ratio $c = \delta_x/\delta_t = 1$ is the lattice velocity. The lattice speed of sound c_s is used for determining the fluid pressure by $p = \rho c_s^2$, and the lattice relaxation time is $\tau = \lambda/\delta_t$. The kinematic viscosity is derived from the relaxation time by the following formula:

$$\nu = (\tau - 0.5)c_s^2\delta_t \tag{2.5}$$

The equilibrium distribution function of Eq. (2.4) is calculated as follows:

$$f_i^{eq} = \rho\omega_i\left[1 + \frac{3}{c^2}\mathbf{c}_i \cdot \mathbf{u} + \frac{9}{2c^4}(\mathbf{c}_i \cdot \mathbf{u})^2 - \frac{3}{2c^2}\mathbf{u} \cdot \mathbf{u}\right] \tag{2.6}$$

where $\mathbf{c}_i = \mathbf{e}_i/\delta_t$ is the lattice velocity in the ith direction, ω_i are the weighting constants for the various lattice links in D2Q9:

$$\omega_i = \lfloor 4/9; 1/36; 1/9; 1/36; 1/9; 1/36; 1/9; 1/36; 1/9 \rfloor \tag{2.7}$$

$\mathbf{u}$ and ρ are the macroscopic velocity and density, respectively. The macroscopic density and momentum are calculated from the distribution function as follows:

$$\rho = \sum_{i=0}^{Q-1} f_i = \sum_{i=0}^{Q-1} f_i^{eq} \tag{2.8}$$

$$\rho\mathbf{u} = \sum_{i=1}^{Q-1} \mathbf{c}_i f_i = \sum_{i=1}^{Q-1} \mathbf{c}_i f_i^{eq} \tag{2.9}$$

where Q depends on the dimension and the type of the LBM model.

Through a Chapman-Enskog expansion in the low frequency, long wavelength limits, and at low Mach number, the LBM can recover the Navier–Stokes equations to a second-order accuracy if the right choice of the equilibrium distribution function is used (Chen et al. 1992; Guo et al. 2000; Latt 2007).

2.1.2 *The Multicomponent LBM*

Two famous multicomponent LBM schemes are reviewed in here: the Gunstensen model (Gunstensen et al. 1991) and the particle-interaction-potential model (Shan and Chen 1993, 1994).

2.1.2.1 The Gunstensen Model

The Gunstensen model identifies a red and a blue momentum distribution functions as $R_i(x, t)$ and $B_i(\boldsymbol{x}, t)$, where x and t are the nodal position and time, respectively. The total momentum distribution function is the sum of the two functions (Gunstensen et al. 1991):

$$f_i(\boldsymbol{x},t) = R_i(\boldsymbol{x},t) + B_i(\boldsymbol{x},t) \tag{2.10}$$

The main difference between the two-component and the single component LBM is the modification of the collision rules in order to induce surface tension and segregate the two immiscible fluids. This is achieved by applying two-step collision rules (Gunstensen et al. 1991; Halliday et al. 2005, 2006, 2007; Hollis et al. 2007; Reis and Philip 2007). The main streaming and collision function is expressed as follows:

$$f_i(\boldsymbol{x} + \boldsymbol{c}_i\delta_t, t + \delta_t) = f_i(\boldsymbol{x},t) - \frac{1}{\tau}\{f_i(\boldsymbol{x},t) - f_i^{eq}(\rho, \rho\boldsymbol{u})\} + \phi_i(\boldsymbol{x}) \tag{2.11}$$

where $\boldsymbol{c}_i$ is the lattice velocity vector in the ith direction shown in Fig. 2.1, τ is the lattice relaxation time, and $\phi_i(\boldsymbol{x})$ is a source term used to induce an interfacial pressure step in the fluid mixture as per Lishchuk's interface method (Lishchuk et al. 2003, 2008). The source term can also enclose a force in the flow direction, which causes fluid circulation. To define the interface between the two fluids, a phase field is described as follows (Halliday et al. 2007):

$$\rho^N(\boldsymbol{x},t) = \frac{R(\boldsymbol{x},t) - B(\boldsymbol{x},t)}{R(\boldsymbol{x},t) + B(\boldsymbol{x},t)} \tag{2.12}$$

where N indicates the direction normal to the interface and the nodal red and blue densities are expressed by the following:

$$R(\boldsymbol{x},t) = \sum_{0}^{Q-1} R_i(\boldsymbol{x},t) \tag{2.13}$$

$$B(\boldsymbol{x},t) = \sum_{0}^{Q-1} B_i(\boldsymbol{x},t) \tag{2.14}$$

The two fluids can have different viscosities. This requires the use of different relaxation times in Eq. (2.5). The interface is made of a fluid mix; therefore its viscosity is determined by the following equation (Dupin et al. 2003):

$$\nu_{eff} = (\tau_{eff} - 0.5)c_s^2\delta_t = \left(\frac{R}{R+B}\right)\nu_R + \left(\frac{B}{R+B}\right)\nu_B \tag{2.15}$$

Lishchuk's interface method is implemented to create a pressure step across the interface. The following surface tension force $\boldsymbol{F}(\boldsymbol{x})$ is used (Lishchuk et al. 2008):

$$\boldsymbol{F}(\boldsymbol{x}) = -\frac{1}{2}\alpha k \nabla \rho^N \tag{2.16}$$

where $\nabla \rho^N = 0$ for a constant phase field. This means that this force is only applicable on the interface. α is a surface tension parameter and k is the curvature of the interface. k is obtained from the surface gradients by solving the following equation using the finite difference method (Lishchuk et al. 2003):

$$k = n_x n_y \left(\frac{\partial n_x}{\partial y} + \frac{\partial n_y}{\partial x} \right) - n_x^2 \frac{\partial n_y}{\partial y} - n_y^2 \frac{\partial n_x}{\partial x} \tag{2.17}$$

where n_x, n_y are the x and y components of the interface normal vector $\boldsymbol{n} \equiv -\nabla \rho^N / |\nabla \rho^N|$.

The following is pseudocode for calculating the derivative of a function in LBM

```
DOFOR x = 0, XLENGTH
  DOFOR y = 0, YLENGTH
    DOFOR i = 1, Q
      n_x = next_x_x[i], n_y = next_y_y[i]
      nn_x = next_x_nx[i], nn_y = next_y_ny[i]
     nnn_x = next_x_nnx[i], nnn_y = next_y_nny[i]
     XDer_func_x,y = XDer_func_x,y + weighting_factor * ( 4.5 * func_n_x,n_y - 0.9 *
     func_nn_x,nn_y + 0.1 * func_nnn_x,nnn_y ) * cx[i]
     YDer_func_x,y = YDer_func_x,y + weighting_factor * ( 4.5 * func_n_x,n_y - 0.9 *
     func_nn_x,nn_y + 0.1 * func_nnn_x,nnn_y ) * cy[i]
    END DO
  END DO
END DO
```

$\boldsymbol{F}(\boldsymbol{x})$ from Eq. (2.16) is used to correct the velocity by Guo's methodology (Guo et al. 2002; Dupin et al. 2003) as follows:

$$\boldsymbol{u}^* = \frac{1}{\rho} \left[\sum_{i=1}^{Q-1} f_i \boldsymbol{c}_i + \frac{1}{2} \boldsymbol{F}(\boldsymbol{x}) \right] \tag{2.18}$$

The relation between the macroscopic and a spatially varying lattice source term is by the following:

$$\phi_i(\boldsymbol{x}) = \omega_i\left(1-\frac{1}{2\tau}\right)[3(\boldsymbol{c}_i-\boldsymbol{u}^*)+9(\boldsymbol{c}_i\cdot\boldsymbol{u}^*)c_i]\cdot\boldsymbol{F}(\boldsymbol{x}) \tag{2.19}$$

where $\boldsymbol{u}$* is the corrected velocity from Eq. (2.18). For constant body force this relationship is expressed by the following equation (Halliday et al. 2007):

$$\phi_i = \omega_i\frac{1}{k_2}\boldsymbol{F}\cdot\boldsymbol{c}_i \tag{2.20}$$

where $k_2 = 1/3$ and $\boldsymbol{F}$ is a constant macroscopic force such as a body force. The first collision is then applied using the corrected velocities in the calculation of the equilibrium distribution function $f^{eq}(\rho, \rho\boldsymbol{u})$. The second step is the segregation of the two fluids which is achieved by imposing zero diffusivity of one color into the other (Gunstensen et al. 1991). A local color gradient is identified as follows:

$$\boldsymbol{G}(\boldsymbol{x},t) = \sum_{ij} c_i\left(R_j(\boldsymbol{x}+\boldsymbol{c}_i,t) - B_j(\boldsymbol{x}+\boldsymbol{c}_i,t)\right) \tag{2.21}$$

A local color flux is calculated by the following formula:

$$\boldsymbol{J} = \sum_i \boldsymbol{c}_i(R_i(\boldsymbol{x},t) - B_i(\boldsymbol{x},t)) \tag{2.22}$$

The segregation step is achieved by forcing the local color flux to align with the direction of the local color gradient. Thus the colored distribution functions at the interface are redistributed such that $-\boldsymbol{J}\cdot\boldsymbol{G}$ is maximized with the following constraints:

$$\begin{aligned} &\sum_i \widehat{\widehat{R}}_i(\boldsymbol{x},t) = R(\boldsymbol{x},t) \\ &\widehat{\widehat{B}}_i(\boldsymbol{x},t) = \widehat{f}_i(\boldsymbol{x},t) - \widehat{\widehat{R}}_i(\boldsymbol{x},t) \end{aligned} \tag{2.23}$$

where $\widehat{\widehat{B}}_i, \widehat{f}_i, \widehat{\widehat{R}}_i$ are the postcollision postsegregation blue, total, and red distribution functions respectively. The segregation can also be accomplished by a formulaic means as described in the model of Halliday et al. (2007) in accordance with the method of D'Ortona et al. (1995):

$$\begin{aligned} \widehat{\widehat{R}}_i(\boldsymbol{x},t+\delta_t) &= \frac{R}{R+B}\widehat{f}_i(\boldsymbol{x},t+\delta_t) + \beta\frac{RB}{R+B}\omega_i\cos\left(\theta_f-\theta_i\right)|\boldsymbol{c}_i| \\ \widehat{\widehat{B}}_i(\boldsymbol{x},t+\delta_t) &= \frac{B}{R+B}\widehat{f}_i(\boldsymbol{x},t+\delta_t) - \beta\frac{RB}{R+B}\omega_i\cos\left(\theta_f-\theta_i\right)|\boldsymbol{c}_i| \end{aligned} \tag{2.24}$$

where θ_f and θ_i are the polar angle of the color field and the angle of the velocity link, respectively; β is the segregation parameter. After the segregation process the two components propagate separately as follows:

$$R_i(\boldsymbol{x} + \boldsymbol{c}_i\delta_t, t + \delta_t) = \widehat{\widehat{R}}_i(\boldsymbol{x}, t + \delta_t) \tag{2.25}$$

$$B_i(\boldsymbol{x} + \boldsymbol{c}_i\delta_t, t + \delta_t) = \widehat{\widehat{B}}_i(\boldsymbol{x}, t + \delta_t) \tag{2.26}$$

In the presented work the Gunstensen and D'Ortona segregation methods will be used.

Succeeding chapters in this book will need the readers to identify the interface of the droplet. Given below is a pseudocode which helps the readers to identify the droplet interface in the Gunstensen-based LBM.

The following is pseudocode for identifying the interface of a droplet

```
DOFOR x = 0, XLENGTH
  DOFOR y = 0, YLENGTH
   ∇x(ρ^N) = 3*∇ x(ρ^N), ∇y(ρ^N) = 3*∇ y(ρ^N)
   IF ABS(∇(ρ^N) > 2.5E-5
     f_x_x,y = ∇x(ρ^N)/ ABS(∇(ρ^N)
     f_y_x,y = ∇y(ρ^N)/ ABS(∇(ρ^N)
     interface_x,y = 1
   END IF
  END DO
END DO
```

2.1.2.2 The Shan and Chen Model

The Shan and Chen model is suitable for simulating multiphase and multicomponent flows. The model uses the following interaction force between the particles of the same specie (Sukop and Thorne 2006):

$$\boldsymbol{F}(\boldsymbol{x}, t) = -G\psi(\boldsymbol{x}, t)\sum_{i=0}^{Q-1} \omega_i\psi(\boldsymbol{x} + \boldsymbol{c}_i\delta_t, t)\boldsymbol{c}_i \tag{2.27}$$

where G is an interaction strength constant which determines the magnitude of the interaction force, and its sign imposes attraction (negative) or repulsion (positive) between the fluid particles. ψ is a potential function of the density; it must be monotonic and bounded (Shan and Chen 1993, 1994):

$$\psi(\rho) = \psi_0 \exp(-\rho_0/\rho) \tag{2.28}$$

where ψ_0 and ρ_0 are constant used to control the potential function. The equilibrium velocity $\boldsymbol{u}^{eq}$ used for the calculation of the equilibrium distribution function prior to the collision step has to be modified, and the interaction force could be imposed using the method of Buick and Greated (2000) for applying an external force into the LBM:

$$\boldsymbol{u}^{eq} = \boldsymbol{u} + \frac{\tau \boldsymbol{F}}{\rho} \tag{2.29}$$

where $\boldsymbol{u}$ and ρ are the macroscopic velocity and density calculated by Eqs. (2.8) and (2.9). With the modified equilibrium distribution function, a Chapman-Enskog expansion leads to the recovery of the isothermal Navier–Stokes equation (Chen et al. 1992). The pressure–density relationship is governed by the following equation of state (EOS):

$$p = \rho c_s^2 + \frac{G\psi^2(\rho)}{2} c_s^2 \tag{2.30}$$

The multicomponent SC model deals with more than one fluid; therefore a composite macroscopic velocity is used to account for all the constituents of the mixture. Equations (2.8) and (2.9) are thus replaced by the following (Sukop and Thorne 2006):

$$\rho^\sigma = \sum_{i=0}^{Q-1} f_i^\sigma = \sum_{i=0}^{Q-1} f_i^{\sigma,eq} \tag{2.31}$$

$$\boldsymbol{u}' = \frac{\sum_\sigma \frac{1}{\tau^\sigma} \sum_{i=1}^{Q-1} c_i f_i^\sigma}{\sum_\sigma \frac{1}{\tau^\sigma} \rho^\sigma} \tag{2.32}$$

where σ refers to the various mixture contributing components; τ^σ is the individual component relaxation time from which different fluids viscosities can be derived using Eq. (2.5).

The fluid–fluid interaction force is represented by the following equation (Martys and Chen 1996):

$$\boldsymbol{F}^{\sigma}(\boldsymbol{x},t) = -\rho^{\sigma}(\boldsymbol{x},t)\sum_{\sigma'} G_{\sigma\sigma'}\sum_{i=0}^{Q-1}\rho^{\sigma'}(\boldsymbol{x}+\boldsymbol{c}_i\delta_t,t)\boldsymbol{c}_i \tag{2.33}$$

where $\boldsymbol{F}^{\sigma}(x, t)$ is the interaction force exerted on component σ by the neighboring component σ' in the mixture. It is worth mentioning that the magnitude of this force, which creates a pressure jump across the fluid–fluid interface, is dependent on the constant $G_{\sigma\sigma'}$ and it determines the strength of the surface tension.

The fluid–solid interaction force exerted by each fluid component is expressed as follows (Martys and Chen 1996):

$$\boldsymbol{N}^{\sigma}(\boldsymbol{x},t) = -\rho^{\sigma}(\boldsymbol{x},t)\sum_{i=0}^{Q-1} G^{\sigma}_{ads}S(\boldsymbol{x}+\boldsymbol{c}_i\delta_t)\boldsymbol{c}_i \tag{2.34}$$

where $S(x + \boldsymbol{c}_i\delta_t)$ can only have a zero value for neighboring fluid node and one for neighboring solid node respectively. G^{σ}_{ads} determines the interaction strength and it is positive for nonwetting fluid, and negative for wetting fluid. The force due to gravity is incorporated in the model through the following:

$$\boldsymbol{E}^{\sigma}(\boldsymbol{x},t) = \rho^{\sigma}(\boldsymbol{x},t)\boldsymbol{g} \tag{2.35}$$

where $\boldsymbol{g}$ is the gravitational constant. The collision step is calculated by the following equation:

$$\widehat{f}^{\sigma}_i(\boldsymbol{x},t+\delta_t) = f^{\sigma}_i(\boldsymbol{x},t) - \frac{1}{\tau^{\sigma}}\left[f^{\sigma}_i(\boldsymbol{x},t) - f^{\sigma,eq}_i(\rho,\rho\boldsymbol{u}^{\sigma,eq})\right] \tag{2.36}$$

where $\widehat{f}^{\sigma}_i$ refers to postcollision distribution functions for the various fluids, and τ^{σ} are their corresponding relaxation times. The equilibrium functions for the constituent fluids are calculated by Eq. (2.6) using the following equilibrium velocities:

$$\boldsymbol{u}^{\sigma,eq} = \boldsymbol{u}' + \frac{\tau^{\sigma}(\boldsymbol{F}^{\sigma} + \boldsymbol{N}^{\sigma} + \boldsymbol{E}^{\sigma})}{\rho^{\sigma}} \tag{2.37}$$

The streaming step is executed for the various fluids using the following equation:

$$f^{\sigma}_i(\boldsymbol{x}+\boldsymbol{c}_i\delta_t,t+\delta_t) = \widehat{f}^{\sigma}_i(\boldsymbol{x},t+\delta_t) \tag{2.38}$$

This is followed by calculating the macroscopic observables using Eqs. (2.31) and (2.32).

2.1.3 Grid Refinement Methods

To extend the applicability of the LBM to a variety of problems including those with turbulent flows, flows in complex geometries like porous media and special boundary shapes, several models were introduced to improve the LBM results quality and to save computational time (Filippova and Hanel 1998; He and Doolen 1997; He et al. 1996; Huang et al. 2007a, b; Inamuro et al. 2005; Kandhai et al. 2000; Li et al. 2005; Lin and Lai 2000; Liu et al. 2009; Shu et al. 2001; van der Sman 2004; Yu and Girimaji 2006; Yu et al. 2002). These models can be classified either by the method used: interpolation, hybrid LBM, and grid refinement, or by the nature of the grid: structured and unstructured gird.

Interpolation method was first used by He et al. (1996), who noticed that the density distribution function is continuous in the physical space; therefore it was possible to define it on nonuniform grid through interpolation. The method was further extended by Shu et al. (2001) and Li et al. (2005) who used Taylor series expansion and least square to evaluate the distribution function instead of direct interpolation. Inamuro et al. (2005) used local time step on nonuniform grid to accelerate the solution since each grid point had its own time step based on the local advection time stability condition.

Hybrid LBM for unstructured grid combined LBM with traditional CFD tools like finite difference, finite volume, and finite element. Hybrid LBM benefited from the LBM stability, resulting from the use of the particle instead of macroscopic velocity. This guaranteed the satisfaction of the Courant-Friedrichs-Lewy (CFL) stability condition. Hybrid LBM gained also the accuracy and efficiency of the traditional CFD tools (Kandhai et al. 2000; Huang et al. 2007a, b).

An interesting unstructured LBM model was proposed by van der Sman (2003), in which no interpolation was required, since particle velocity in this model was different for different lattice sites. This led to the elimination of the undesired numerical diffusion caused by the interpolation step.

Grid refinement methods worked by locally refining the mesh in parts of the domain characterized with complex geometry and where higher accuracy was required. Filippova and Hanel (1998) introduced the first model which was based on passing data from the postcollision distribution functions between the coarse and fine grids. The transfer of data maintained continuous viscosity and therefore Reynolds number throughout the domain. The model handled very well complex geometries by specially treating curved boundaries. Lin and Lai (2000) proposed a composite block-structured LBM in which a coarse grid covered the whole domain, and only areas of interest were patched with a fine grid blocks. This method did not need time interpolation, because solutions on both grids were at the same time level. Yu et al. (2002) suggested a very efficient multiblock method in which fine and coarse grids did not overlap throughout the fine grid block, and data transfer occurred only at the interface boundary nodes. This method was later expanded to three-dimensional models by Yu and Girimaji (2006).

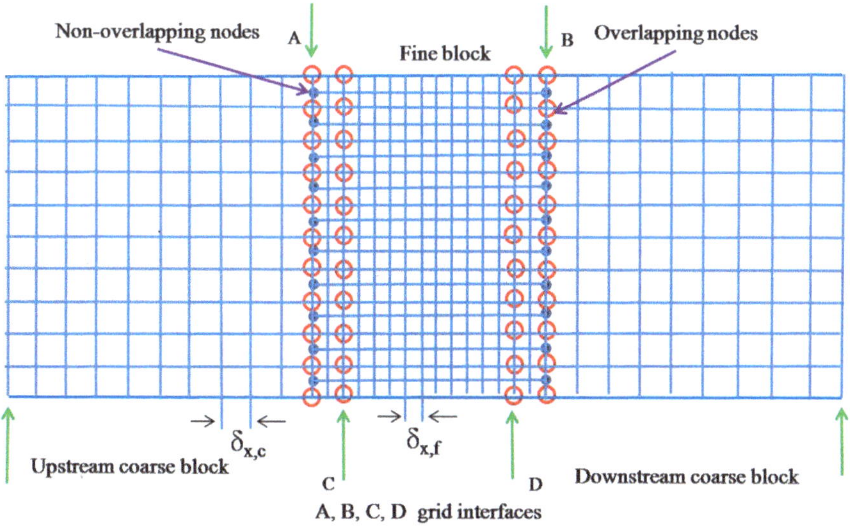

Fig. 2.2 Illustration of the grid interfaces between the fine and the coarse block from the proposed multiblock scheme

Multiphase and multicomponent flows were not as extensively studied with respect to grid refinement as turbulent flows and near-solid boundary phenomena. The peculiarity of the multicomponent flows is due to the movement and deformation of the suspended fluid. This hampers the use of preset grid refinement techniques since the area of interest is not fixed in the domain. Tolke et al. (2006) proved in their Gunstensen-based LBM that the interface was distorted relative to the magnitude of the capillary forces. This was observed when they allowed the fluid interface to pass through the grid interface of different preset fixed grids. They also indicated through a mathematical model that the grid level which could be used in such cases was very restricted. Thus they resorted to the use of an adaptive grid method, in which the physical interface was always discretized on the finest grid level. Ozawa and Tanahashi (2005) presented a model for multiphase flow, with an adaptive unstructured grid. Ozawa and Tanahashi (2005) used cubic interpolation with volume/area coordinates method for the streaming step and moving least-square method for the collision step. The mesh was refined based on a number density threshold using Bisection algorithm.

The grid refinement method of Yu et al. (2002) will be extended here in what will be called the migrating multiblock scheme in order to allow its use for multiphase and multicomponent flows. Yu et al. (2002) proposed the following relationships between the distribution functions of the various blocks shown in Fig. 2.2. The grid spacing ratio is expressed through the following equation:

$$m = \frac{\delta_{x,c}}{\delta_{x,f}} \tag{2.39}$$

The relationship between the relaxation times of the various blocks is expressed as follows:

$$\tau_f = \frac{1}{2} + m(\tau_c - 0.5) \tag{2.40}$$

The exchange of data between the various blocks occurs at the grid interface where fine and coarse nodes overlap through the following:

$$\widehat{f}_i^f = f_i^{eq,c} + \frac{\tau_f - 1}{m(\tau_c - 1)}\left[\widehat{f}_i^c - f_i^{eq,c}\right] \tag{2.41}$$

$$\widehat{f}_i^c = f_i^{eq,f} + m\frac{\tau_c - 1}{(\tau_f - 1)}\left[\widehat{f}_i^f - f_i^{eq,f}\right] \tag{2.42}$$

where the postcollision distribution function of the fine grid is $\widehat{f}_i^f$, while $\widehat{f}_i^c$ is the postcollision distribution function of the coarse grid. A symmetrical cubic spline interpolation was used for spatial interpolation of the postcollision distribution functions on the fine block boundary.

A three-point Lagrangian formula was implemented in order to synchronize the time steps in the various blocks (Yu et al. 2002):

$$\widehat{f}_i^f(t) = \sum_{p=1}^{3}\left[\widehat{f}_i^f(t_p)\prod_{q=1,p\neq q}^{3}\frac{t - t_q}{t_p - t_q}\right] \tag{2.43}$$

where p, q are positive integers ranging from 1 to 3.

2.2 Colloidal Studies

2.2.1 *Surfactant-Laden Droplets*

Surfactant-covered droplets were investigated experimentally, analytically, and numerically. Sundararaj and Macosko (1995) studied the role of compatibilizers in stabilizing the morphology of some polymer blends as a result of suppressing the coalescence of the dispersed phase. Williams et al. (1997) investigated the effects of protein emulsifiers on the breakup of a single aqueous drop in shear flow, and found out that at high emulsifier concentrations the drop size was two orders of magnitude smaller than the expected size from its equilibrium interfacial tension. Lyu et al. (2002) used block copolymer to reduce the particle size in polymer blends and attributed this process to the steric repulsion which depended on the surfactant molecular weight. Hu and Lips (2003) delineated the individual effects of the dilution, the Marangoni stress and the tip stretching on surfactant covered mother drop by measuring the interfacial tension of the subdivided generations of daughter drops. Almatroushi and Borhan (2004) examined the effect of surfactants on the buoyancy of bubbles and viscous drops in a bounded domain.

Vlahovska (2003) and Vlahovska et al. (2005) developed analytical solutions for the small perturbation of the surfactant distribution, which influenced the drop evolution in linear flows. The solutions for the system were formulated as a nonlinear matrix equation after expanding the velocity, surfactant concentration, and the drop shape in spherical harmonics.

Milliken et al. (1993) studied the effect of dilute insoluble surfactant on the deformation of a drop in uniaxial extensional flow using boundary integral technique to describe the motion of the drop interface, and finite difference scheme for the mass transfer at the interface. Li and Pozridikis (1997) used similar numerical approach with a linear surfactant equation of state to study the transient deformation of spherical drop with viscosity ratio of one with respect to the matrix. Eggleton et al. (2001) studied tip streaming and drop breakup dependence on the surfactant initial coverage in linear extensional flow. Their model used boundary integral formulation for the Stokes equations, Runge–Kutta method for the interface time evolution, and finite difference for the mass balance equation. Drumright-Clarke (2002) and Drumright-Clarke and Renardy (2004) used the volume of fluid method to track the interface, the projection method to solve the Navier–Stokes equations, and the continuum method for the surface tension to model the effects of dilute insoluble surfactant on a drop in strong shear flow. Kruijt-Stegeman et al. (2004) used a finite element method to study the transient deformation of drops in supercritical elongational flow and the breakup of elongated drops in quiescent medium with low surfactant coverage. The surface tension was incorporated into the finite element as a volume force. Lai et al. (2008) proposed an immersed boundary method for modeling fluid interfaces with insoluble surfactant in 2D geometries. A symmetric discretization for the surfactant concentration was employed to insure surfactant mass conservation numerically. Finally van der Graaf (2006) and van der Sman and van der Graaf (2006) used a free energy-based LBM to develop a diffuse model for studying the adsorption of surfactant onto flat and drops interfaces. The model was tested briefly in 2D linear shear and uniform flow fields to show its applicability when coupled to the hydrodynamics.

The following studies (Lyu et al. 2002; Jeon and Macosko 2003; Milliken and Leal 1991; Hu and lips 2003; Cheng et al. 2005; Sundararaj and Macosko 1995; Kleshchanok and Lang 2007) provide a good understanding of the physical interaction and deformation of droplets during their formation and breakup. The following facts are selected due to their relevance to the subject of this work:

- Surfactants reduce the surface tension and therefore enhance the deformability of the droplets by simply increasing their capillary number. The capillary number is the ratio of the viscous stress and the Laplace pressure $Ca = \frac{\mu R \dot{\gamma}}{\alpha}$, where μ is the viscosity of the suspending fluid, R is the droplet radius, $\dot{\gamma}$ is the shear rate, and α is the surface tension. The surfactant role is affected by three additional mechanisms namely, surface dilution which is due to the increase of

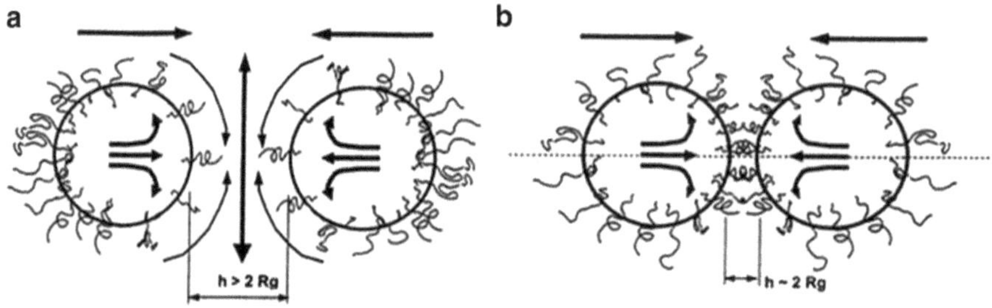

Fig. 2.3 Two mechanisms for suppressing coalescence are presented here. (**a**) Surfactant concentration gradient. (**b**) Steric repulsion (Lyu et al. 2002)

the droplet area during deformation, tip-stretching which is caused by the convection of the surfactant towards the droplet tips due the bulk flow, and the Marangoni stresses originating from the gradient in the surfactant concentration along the interface.

- Surfactants suppress coalescence leading to stabilization of the mixture. There are two different theories for the explanation of the suppression of coalescence shown in Fig. 2.3; the first (Fig. 2.3a) advocates that the Marangoni stresses increase on the opposing interfaces due to the squeeze of the matrix between the droplets. This retards the local interface velocity, thus slows down the film drainage and consequently prevents coalescence. The second (Fig. 2.3b) claims that the suppression of coalescence is due to steric repulsive force generated by the compression of the surfactant layers, which are attached to the surfaces of two approaching droplets, and that steric repulsion is a surfactant molecular weight dependent force (Lyu et al.; 2002). Experimental works (Lyu et al. 2002; Kleshchanok and Lang 2007) showed that block copolymers with higher molecular weights have greater tendency to suppress coalescence.
- Surfactants could also lead to aggregation or what is called adsorption flocculation. Two models are distinguished to explain this phenomenon. The bridging model (Lyu et al. 2002) is based on the principle that the adsorbed macromolecules on adjacent droplets surfaces create a bridging force which exceeds the forces of disaggregation, hence enhancing flocculation. The depletion model (Neu and Meiselman 2002) proposes that droplets aggregation is due to an exclusion of the macromolecules from the droplet interface which creates an osmotic pressure difference favoring aggregation.
- It is generally agreed that the following interaction forces are at work between two approaching surfactant covered droplets; steric repulsive interaction, electrostatic repulsive interaction, the van der Waals attractive interaction, and the bridging or depletion interactions. The resultant of these combined forces determines whether coalescence will occur or not. This force description does not deviate from the explanation given by Lipowsky and Sakmann (1995) on the nature of force interactions between two biological membranes (with the exception of the presence of a repulsive hydration force). These forces decay

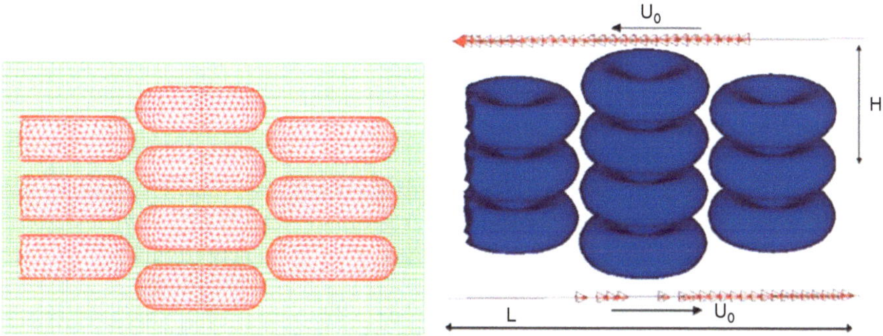

Fig. 2.4 Illustration of the domain for the simple shear flow used to calculate effective viscosity (Liu and Liu 2006)

depending on the distance between the droplets either by an exponential or by inverse square laws.

- Surfactants generally reduce significantly the particles' terminal velocity below the classical Hadamard-Rybszynski prediction in the spherical region of the shape regime; however in other shape regions the particle retardation due to surfactants is less effective (Tasoglu et al. 2008).

2.2.2 *Colloids Rheology*

Colloids are non-Newtonian fluids, i.e., their relative viscosity is dependent on the shear rate in the bulk flow. Two major approaches for the numerical study of colloids rheology are found in the literature: indirect and direct methods. The indirect methods are those in which the rheological behavior of the suspension is usually incorporated in a single component model through the integration of some empirical formulae (Artoli 2003; Ouared and Chopard 2005; Chandran et al. 2006). The most famous of these is the Cason's $\sqrt{\mu|\dot{\gamma}|} = \sqrt{\sigma} - \sqrt{\sigma_y}$ which was used by Ouared and Chopard (2005) where σ is the shear stress, σ_y is yield stress, the power law $\sigma = \mu\dot{\gamma}^n$, and the Bingham plastic $\sigma = \sigma_y + \mu_b\dot{\gamma}$, (Chandran et al. 2006) where μ_b is the Bingham viscosity coefficient.

The direct methods are based on rather measuring the pressure drop Δp along the length of a channel and the flow rate Q in Poiseuille flow to deduce the apparent viscosity by $\mu_{app} = \frac{(\Delta p/\Delta L)\pi r^4}{8Q}$ (Sun and Munn 2005, 2006), or measuring the shearing force per unit width F in Couette flows as in Fig. 2.4, and calculate the effective viscosity $\mu_{eff} = \frac{HF}{LU_0}$ (Liu and Liu 2006), where H is half the channel width, L is the channel length, and U_0 is the wall velocity.

2.2.3 Red Blood Cell Deformability

Red blood cell (RBC) deformability is the most important physical property to study when analyzing blood flow inside the capillaries. Due to this extreme property RBCs are capable of streaming through vessel with diameters half of their size. Factors affecting RBC deformability are many. These include some serious illnesses like malaria, diabetes, atherosclerosis, and others. A list of some facts related to RBC deformation mechanisms was extracted from the literature (Baskurt and Meisleman 2003; Braasch 1971; Gedde et al. 1995; Keller et al. 1998):

- In the microvasculature the velocity of the RBC is dependent on the level of its rigidity and shape.
- Deformability of the RBC generally helps reduce the blood viscosity both in large and small vessels.
- Severe shocks, burns, and some snake bites rigidify the RBC membrane.
- The deformability of the RBC is influenced by the surface area to volume ratio of the cell. Therefore swelling of the RBC due to a reduction in the blood osmolality decreases its deformability.
- Shrinking of the RBC (Crenated cells) due to increase in the osmolarity of the blood. This changes the internal cell fluid viscosity causing higher RBC rigidity.
- The availability of metabolic energy, i.e., adenosine triphosphate, which is very essential for the functioning of the cation pump in the RBC membrane. This pump maintains all required active substance exchange like intercellular cation and water, thereby maintaining the RBC surface area to volume ratio. This is also important for the reduction of cytosolic calcium concentration, which excesses rigidify the RBC cytoskeleton. The lack of metabolic energy often happens in the events of glucose shortage in the RBC microenvironment.
- Oxygen-free radicals associated with an ischemia–reperfusion injury generate superoxide anions inside the RBC which decrease its deformability.
- Polymorph nuclear leukocytes are activated during injury or inflammation resulting in increased level of secretory activities which cause reduced RBC deformability.

Describing the RBC deformability, Baskurt and Meisleman (2003) claimed that the alteration in the composition of the lipid bilayer had a minor role in the mechanical behavior of the membrane, and that the cytoskeleton protein constituents had a major role in this process. The cytoskeleton is believed to be susceptible to chemical reactions which increase the cross-linkage among membrane skeletal proteins thereby reducing the cell deformability (Fig. 2.5).

The role of the lipid bilayer alteration was more appreciated (Gedde et al. 1995; Braasch 1971). Two possible mechanisms were explained by Gedde et al. (1995).

- The perturbation of the distribution of the phosphatidylcholine in the outer leaflet and the phosphatidylserine in the inner leaflet can cause a change in the membrane curvature.

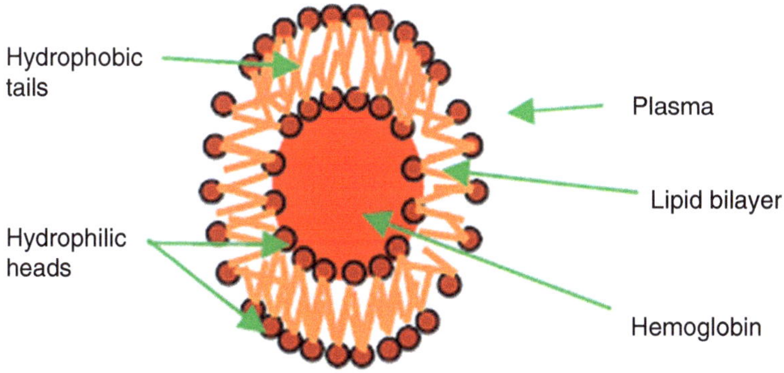

Fig. 2.5 Simplified representation of the blood cell as Liposome, justified by the assumption that the membrane liquefies under pressure in the microvasculature

- The change in the electrostatic interactions of the lipid head groups has a major influence on the membrane shape. Both mechanisms were tested by the titration of the cell pH (logarithmic measure of hydrogen ion concentration).

Braasch (1971) explained the Norris surface tension hypothesis and the role of surface active substances on the RBC shape and deformation. Braasch gave the following evidence in support of this hypothesis:

- The sphering ability of surface active substances, like saponin, free fatty acids, bile, lycolecithin, and some snake venoms.
- Crenated cells are induced by anionic and nonionized compounds, while cup-shape cells are by cationic compounds. Chlorpromasine stretches biological membranes.
- Incubation of RBC in plasma at 37° causes sphering of the cell due to esterification of the cholesterol in the plasma which was replaced by the cholesterol of the RBCs.
- Sphering and crenation of the RBC occurrence after severe shocks which was caused by the effects of adrenaline and catecholamine in the blood.

The influence of the lipid bilayer on the deformation of the RBC was further supported by the results of the experiment Keller et al (1998), which indicated that the RBC lipids formed immiscible fluid below pressures of 21 dyne/cm for the inner leaflet, and 29 dyne/cm for the outer leaflet. At higher pressures the lipids mixed together and formed a homogeneous liquid.

Chapter 3
Accelerated Lattice Boltzmann Method

Abstract The LBM is a powerful numerical fluid solver especially in the areas of multicomponent and multiphase mixtures. However the LBM uses the particle velocity for the determination of the model time step required by the Courant-Friedrichs-Lewy (CFL) stability condition. This degrades the LBM efficiency, when compared with standard CFD solvers which use instead the macroscopic velocity for the CFL requirements. To increases the efficiency of the LBM and especially for multiphase flow simulations, the migrating multiblock (MMB) was introduced.

This chapter presents an extension scheme for the application of the single phase multiblock lattice Boltzmann method (LBM) to the multiphase Gunstensen and Shan and Chen models. The grid is refined for part of the domain where a fluid–fluid interface evolves, and the refined grid is free to migrate with the suspended phase in the flow direction. The method is applicable to single and multiphase flows in 2D and 3D domains. This accelerated scheme was demonstrated by simulating a benchmark single phase flow around a 2D asymmetrically placed cylinder in a channel and for investigating the shear lift of 2D neutrally buoyant bubble in a parabolic flow. The method was also used for simulating cases of 3D rising bubbles in infinite medium, in which the model results for the bubble terminal velocity were in good agreement with a semianalytical solution, and the produced shapes fitted very well in an experimental shape regime map. The MMB was also used for the study of the sedimentation of droplet onto horizontal surfaces with different wetting characteristics of wall surface, which led to the identification of three different regimes of droplet shapes. It was shown that the process of different regimes can be controlled by changing the wetting characteristics of surface.

Electronic supplementary material Supplementary material is available in the online version of this chapter at http://dx.doi.org/10.1007/978-1-4899-7402-0_3. Videos can also be accessed at http://www.springerimages.com/videos/978-1-4899-7401-3

H. Farhat et al., *Accelerated Lattice Boltzmann Model for Colloidal Suspensions: Rheology and Interface Morphology*, DOI 10.1007/978-1-4899-7402-0_3,

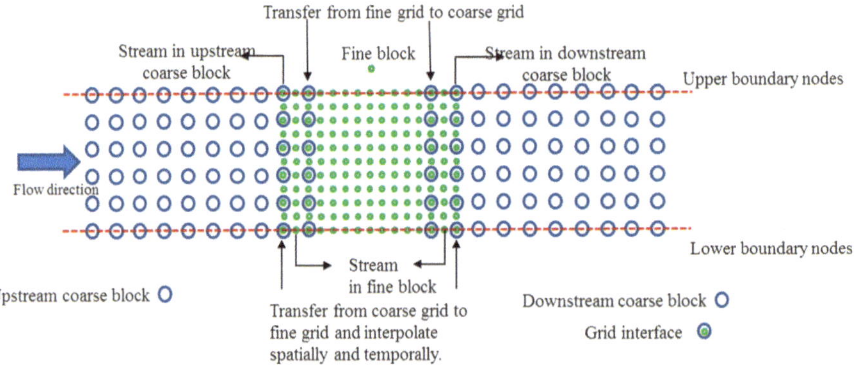

Fig. 3.1 Illustration of the standard multiblock LBM domain as it pertains to the single component flow simulation of this work

3.1 Migrating Multiblock Schemes for the D2Q9 LBM

The present work's objective is to provide a simple algorithm, aiming at saving considerable computational time in simulations where local grid refinement is required, and especially applicable to multiphase and multicomponent flows with highly deformable interface. To avoid the difficulties faced by Tolke et al. (2006) and to maintain a relatively simple approach using standard structured grid LBM, a Gunstensen-based model combined with the multiblock method of Yu et al. (2002) is presented here. The difference in this model lies in that a fine grid block covers the entire fluid interface and migrates with it, so that the physical interface does not cross the grid interface. This is performed by tracking the mass center or the average velocity of the suspended fluid, which acts as a trigger to impose node type exchange at the grid interfaces in a way that does not alter the physical properties of the various fluids. The node type exchange occurs without time lag during the propagation step in the coarse block. The grid interface is always imposed where a single phase exists.

3.1.1 The Migrating Multiblock Algorithms

3.1.1.1 Single Component Algorithm for 2D Geometries

The following is a brief description of Yu et al. (2002) multiblock LBM tailored for this work domain, in which the width is much smaller than the length. The domain shown in Fig. 3.1 consists of three blocks: an upstream coarse block, a fine block, and a downstream coarse block.

The ratio of the lattice spacing *m* between the fine and coarse blocks is defined by Eq. (2.39). To maintain the same viscosity, and therefore the same Reynolds number in the various blocks, the relaxation times have to satisfy Eq. (2.40).

Each grid interface consists of overlapping two sets of coarse and fine nodes with one additional set of fine nodes filling in the gap as shown in Fig. 3.1. The transfer of the postcollision distribution functions $\widehat{f}_i^f \leftrightarrows \widehat{f}_i^c$ between the different grids occurs before the streaming step. To maintain the same lattice velocity ($\delta_x^c/\delta_t^c = \delta_x^f/\delta_t^f$) between the various grids, the time step ratio is the same as the spatial ratio ($m = \delta_x^c/\delta_x^f = \delta_t^c/\delta_t^f$). After one collision step in the coarse block a transfer of data $\left(\widehat{f}_i^f \leftarrow \widehat{f}_i^c\right)$ is required at the indicated locations in Fig. 3.1 by Eq. (2.41). After m collision steps in the fine grid a transfer of data $\left(\widehat{f}_i^f \rightarrow \widehat{f}_i^c\right)$ is required at the indicated locations in Fig. 3.1 by Eq. (2.42). A symmetric cubic spline interpolation is required for calculating $\widehat{f}_i^f$ on those fine nodes, which do not overlap with the coarse nodes at the grid interface. This is done by the following formula (Rorres and Howard 1984):

$$\widehat{f}_j^f(x) = a_j\left(x - x_j\right)^3 + b_j\left(x - x_j\right)^2 + c_j\left(x - x_j\right) + d_j \tag{3.1}$$

The coefficients in Eq. (3.1) are calculated as follows:

$$\begin{aligned} a_j &= M_{j+1} - \frac{M_j}{6h} \\ b_j &= \frac{M_j}{2} \\ c_j &= \left(f_{j+1}^f - f_j^f\right) - \left(\frac{M_{j+1} + 2M_j}{6}\right)h \\ d_j &= f_j^f \end{aligned} \tag{3.2}$$

where M_j are second-order derivatives of the function $\widehat{f}_j^f$ and $h = x_j - x_{j-1}$ measured in the coarse block. The M_j functions are calculated by solving a matrix equation, which leads to a tridiagonal coefficients matrix suitable for the Thomas algorithm, and the natural spline condition is stipulated where $M_0 = M_n = 0$.

A three-point Lagrangian interpolation scheme is used to calculate the postcollision distribution function on the grid intersection by Eq. (2.43). This leads to the following relation for the temporal interpolation with for example a spacing ratio $m = 4$ and time measured in coarse steps:

$$\begin{aligned} \widehat{f}_i^f(t_{+0.25}) &= -0.09375\widehat{f}_i^f(t_{-1}) + 0.4375\widehat{f}_i^f(t) + 0.65625\widehat{f}_i^f(t_{+1}) \\ \widehat{f}_i^f(t_{+0.5}) &= -0.125\widehat{f}_i^f(t_{-1}) + 0.75\widehat{f}_i^f(t) + 0.375\widehat{f}_i^f(t_{+1}) \\ \widehat{f}_i^f(t_{+0.75}) &= -0.09375\widehat{f}_i^f(t_{-1}) + 0.9375\widehat{f}_i^f(t) + 0.15625\widehat{f}_i^f(t_{+1}) \end{aligned} \tag{3.3}$$

For simplicity a ratio $m = 2$ was used throughout this work. This required the utilization of only $\widehat{f}_i^f(t_{+0.5})$ from Eq. (3.3) for the temporal interpolation.

The following is pseudocode which defines the skeleton of main function for a multiblock algorithm where m = 2 and droplet moving in positive x-direction

```
Main
{
 Define Parameters
Subroutine boundary_conditions
DOFOR timestep_coarse = 0, T, 2
  Subroutine Migration_fineblock
  Subroutine Propogate_coarseblock1
  Subroutine Propogate_coarseblock2
  Subroutine Collision_coarseblock1
  Subroutine Collision_coarseblock2
  DOFOR timestep_fine = 0,1
    Subroutine Propogation_fineblock
     IF timestep_fine = 0
      Subroutine transfer_coarse1_fine
      Subroutine transfer_coarse2_fine
      Subroutine spatial_interpolate_block_interface1
      Subroutine spatial_interpolate_block_interface2
    END IF
    Subroutine Collision_fineblock
     IF timestep_fine = 0
      Subroutine time_interpolate_block_interface1
      Subroutine time_interpolate_block_interface2
    END IF
  END DO
  Subroutine transfer_fine_coarse1
  Subroutine transfer_fine_coarse2
END DO
```

3.1.1.2 Gunstensen Multicomponent Algorithm for 2D Geometries

To implement the multiblock concept on the Gunstensen model care should be taken of the collision step which involves the sum $f_i^f(\boldsymbol{x}, t)$ of the two distribution functions $R_i(\boldsymbol{x}, t)$ and $B_i(\boldsymbol{x}, t)$ as it was expressed in Eqs. (2.10) and (2.11). Therefore the sum postcollision distribution function $\widehat{f}_i^f(\boldsymbol{x}, t+\delta_t)$ should be used in Eqs. (2.41) and (2.42) for the required transfers at the grid interfaces. Since the streaming step in the Gunstensen model occurs with separate postcollision postsegregation distribution functions $\widehat{\widehat{R}}_i(\boldsymbol{x}, t)$ and $\widehat{\widehat{B}}_i(\boldsymbol{x}, t)$, a transfer of the grid interface information from the sum function $\widehat{f}_i^f(\boldsymbol{x}, t+\delta_t)$ to the suspending component function $\widehat{\widehat{B}}_i(\boldsymbol{x}, t+\delta_t)$ is necessary before streaming. This is to ensure that the transfer of information at the interface between the different grids is propagated through the function $\widehat{\widehat{B}}_i(\boldsymbol{x}, t+\delta_t)$ into the fine block. This transfer is not required for the function $\widehat{\widehat{R}}_i(\boldsymbol{x}, t)$ since the physical interface in the presented model does not cross the grid interface contrary to the experiment of Tolke et al. (2006), and the exchange of information from the various grids is done only at the single phase interface nodes.

The migrating multiblock method's main feature is the exchange of node type at the grid interfaces. For the fluid–fluid interface to be constantly covered by a fine grid while moving, an exchange of boundary coarse nodes with fine nodes downstream of the fluid interface, and alternatively an exchange of fine boundary nodes with coarse nodes upstream of the interface are needed as shown in Fig. 3.2. The node type exchange occurs when the distance travelled by the suspended fluid mass center exceeds one coarse lattice spacing in the flow direction. This exchange happens during one coarse time step and it starts with the streaming in the coarse block. Here a distinction should be drawn between the two coarse blocks. The propagation step should start in the downstream coarse block first, because after propagation the coarse distribution function $f_i^{c,d}(\boldsymbol{x}, t)$ is set to zero at the location indicated as old diminishing coarse nodes in Fig. 3.2, thereby allowing only fine nodes to occupy it. This can be done since the information needed for the propagation has already been passed.

To create two new sets of fine nodes $\widehat{\widehat{B}}_i(\boldsymbol{x}, t)$ on the new grid interface downstream of the suspended fluid, the extrapolation method is used at the location indicated as newly created fine nodes in Fig. 3.2. For spacing ratio $m \gg 2$, a careful selection of the extrapolation scheme should be done in order to minimize any possible numerical diffusion, since the extrapolation method could become less accurate. Meanwhile, during the same coarse time step, at the grid interface the

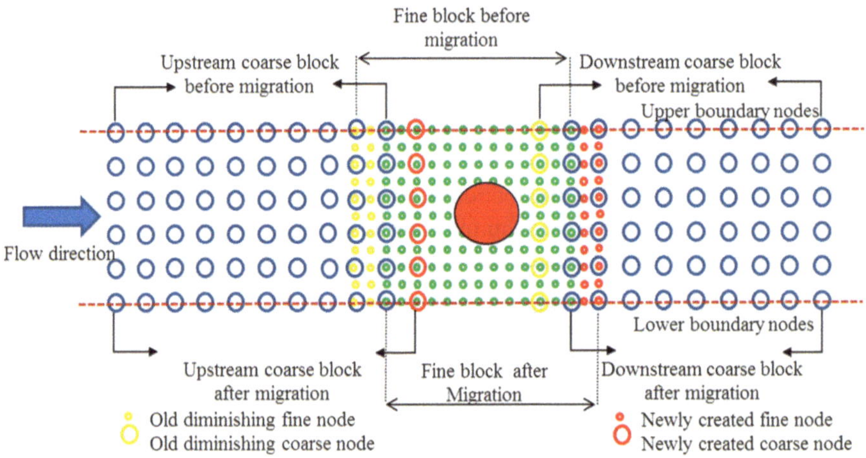

Fig. 3.2 Illustration of the migrating multiblock LBM domain as it pertains to the multiphase flow simulation used in this paper

following functions ($\widehat{f}_i^f\,(\boldsymbol{x}, t)$, $\widehat{\widehat{B}}_i(\boldsymbol{x}, t)$, $\rho^N(\boldsymbol{x}, t)$) are set to zero in the locations designated as old diminishing fine nodes in Fig. 3.2. This will not influence the solution because the information stored in these fine nodes is not required for any subsequent calculations. The propagation in the fine block will not include the vanishing nodes, which position in the domain is now occupied by only coarse nodes. Extrapolation is used again to create a new set of coarse nodes $f_i^{c,u}(\boldsymbol{x}, t)$ at the locations indicated as newly created coarse nodes in Fig. 3.2. This is followed by an immediate transfer of data from the fine node at the new interface to obtain $\widehat{f}_i^{c,u}\,(\boldsymbol{x}, t)$ needed for the propagation in the upstream coarse block. The next step is propagating the upstream coarse block followed by the steps provided in the flow chart of Fig. 3.3 which resemble the steps of the standard multiblock model (see Movie C3.1.1.2—MMPois).

The following is pseudocode for tracking the center of mass of droplet moving along x-axis

```
Subroutine x_0()
 Define parameters
 DOFOR x = 0, XLENGTH
   DOFOR y = 0, YLENGTH
     Wk1 = Wk1 + b_{x,y} * x
     Wk2 = Wk2 + b_{x,y}
   END DO
 END DO
 RETURN VALUE Wk1/Wk2
END Subroutine
```

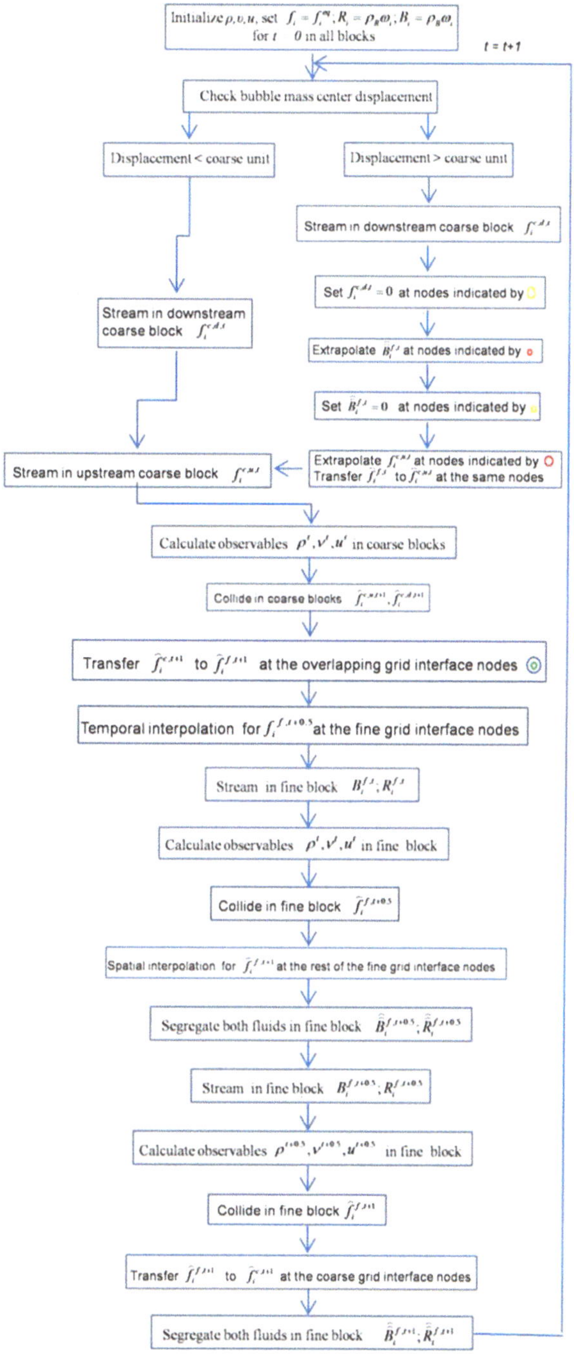

Fig. 3.3 Flow chart for the migrating multiblock LBM for immiscible mixtures

The following is pseudocode for migration of the fine block

```
IF ABS(x_0()–dx)>=2        (dx is initial distance of the droplet center of mass from x = 0)
  DOFOR y = 0, YLENGTH, 2
    DOFOR x = dx – (a-4), dx – (a-5), 2
      DOFOR i = 0, Q
        bcoarse1_{x,y[i]} = 2* bcoarse1_{x-2,y[i]} – bcoarse1_{x-4,y[i]}
      ENDDO
    END DO
  END DO
  DOFOR x = dx+(a-4), dx+(a-5), 2
    DOFOR y = 0, YLENGTH, 2
      DOFOR i = 0, Q
        bcoarse1_{x,y[i]} = f_{x,y[i]} + 6.0 *(bfine_{x,y[i]} – f_{x,y[i]})
      END DO
    END DO
  END DO
  DOFOR x = dx+(a-2), dx+(a-1), 2
    DOFOR y = 0, YLENGTH, 2
      DOFOR i = 0, Q
        bcoarse2_{x,y[i]} = 0
      END DO
    END DO
  END DO
  DOFOR y = 0, YLENGTH
    DOFOR x = dx+(a+1), dx+(a+3)
      DOFOR i = 0, Q
        IF y < 2 OR y > YLENGTH-3
          bfine_{x,y[i]} = Density * weighting_factor
        ELSE
          bfine_{x,y[i]} = 2* bfine_{x-1,y[i]} - bfine_{x-2,y[i]}
        END IF
```

```
          END DO
        END DO
      END DO
      DOFOR x = dx-a, dx-(a-2)
        DOFOR y = 0, YLENGTH
          DOFOR i = 0, Q
            bfine_x,y[i] = f_x,y[i] = 0
          END DO
        END DO
      END DO
      DOFOR x = dx-a, dx-(a-2)
        DOFOR y = 0, YLENGTH
          ρ^N_x,y = 0
        END DO
      END DO
      dx = dx + 2
      Subroutine boundary_conditions
    END IF
```

3.1.2 *Simulation Results and Discussions*

3.1.2.1 Asymmetrically Placed Cylinder in 2D Channel

To investigate the effects of the migrating block on the numerical solution, an unsteady flow around an asymmetrically placed cylinder in 2D channel was simulated using fixed and migrating multiblock schemes simultaneously. The results were compared with some benchmark cases presented by Schafer and Turek (1996).

The center of the cylinder was located at 4.0 radii from the lower wall, 4.2 radii from the upper wall, and 4.0 radii from the inlet as shown in Fig. 3.4. The fine block covered the whole cylinder, and it was 80 by 164 lattice squares. The total number of nodes in both coarse blocks was 32,800. The ratio between the coarse and the fine grids was $m = 2$. The relaxation times for the fine and the coarse grids were

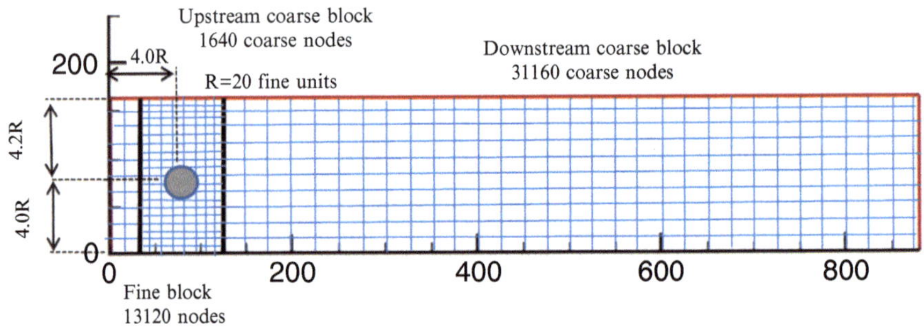

Fig. 3.4 Migrating multiblock LBM domain for the flow around an asymmetrically placed cylinder in a channel, with the cylinder center location expressed as a function of its radius

$\tau_f = 0.58$ and $\tau_c = 0.54$, respectively. The average velocity used for the calculation of the Reynolds number was:

$$\overline{U} = \frac{2}{3} U_{(0,\frac{H}{2},t)} \tag{3.4}$$

where H is the channel height, t is time, and U is the centerline velocity. The average velocity used for this simulation was $\overline{U} = 0.0666$ lattice units per time step, resulting in a Reynolds number Re $= 100$. The extrapolation method was enforced on the outlet boundary, and the bounce back condition was implemented on the top and bottom walls as well as on the cylinder surface. The method of Zou and He (1997) was applied on the inlet of the domain using a parabolic velocity profile which was calculated by the following formula (Schafer and Turek 1996):

$$U_{(0,y,t)} = \frac{4U_{(0,\frac{H}{2},t)} y(H-y)}{H^2} \tag{3.5}$$

Under these conditions an unsteady flow was developed and a periodic vortex shedding was witnessed. Instantaneous streamlines for the fixed block simulation are shown in Fig. 3.5a. The Strouhal number defined as $St = \frac{Df}{\overline{U}}$ where D is the cylinder diameter and f is the frequency of separation (the inverse of the period from peak to peak values of the lift coefficient) was $St = 0.293$. This value matched well with the results given by Schafer and Turek (1996). The same simulation was carried out again with the fine block migrating by one coarse lattice each 5×10^3 coarse time steps. The Strouhal number was calculated as $St = 0.297$. This value agreed well with those given by Schafer and Turek (1996) ($0.295 \leq St \leq 0.305$). A qualitative comparison between the instantaneous streamlines of Fig. 3.5b, with the streamlines of Fig. 3.5a, shows a very marginal difference caused by the moving fine block after five consecutive shifts.

The Strouhal numbers in both simulations were derived using the lift coefficients graph of Fig. 3.6, which was plotted together with the drag coefficients between

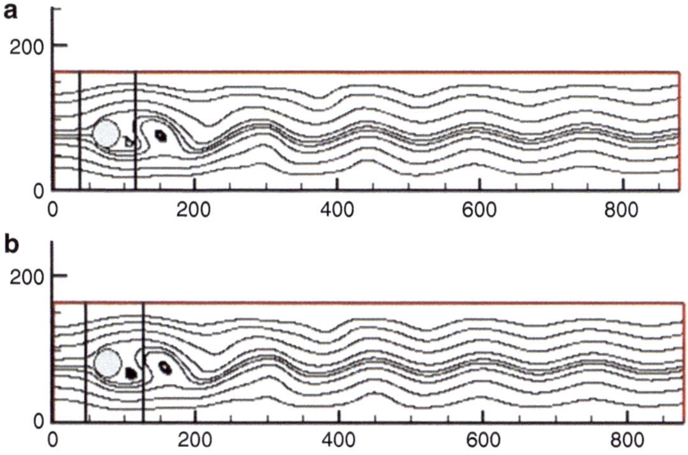

Fig. 3.5 Instantaneous streamlines of a 2D channel flow over an asymmetrically placed cylinder with Re $=$ 100 time step 2.9×10^4 measured in coarse time units. (**a**) Fixed multiblock in which the fine block is static having a center coinciding with the cylinder center, (**b**) migrating multiblock in which the fine block migrated in the direction of the flow by one coarse space unit each 5.0×10^3 coarse time steps and having its center advanced by ten fine space units in the flow direction with respect to the cylinder center

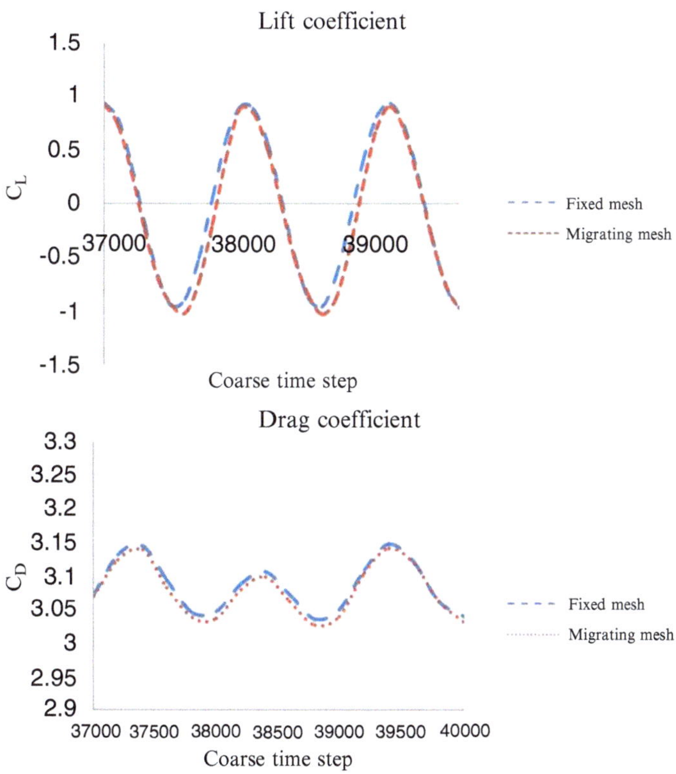

Fig. 3.6 Lift and drag coefficients for fixed and migrating multiblock cases, calculated for results taken between coarse time steps 3.7×10^4 and 4.0×10^4. A comparison of the two cases indicates that the block migration altered the results just marginally

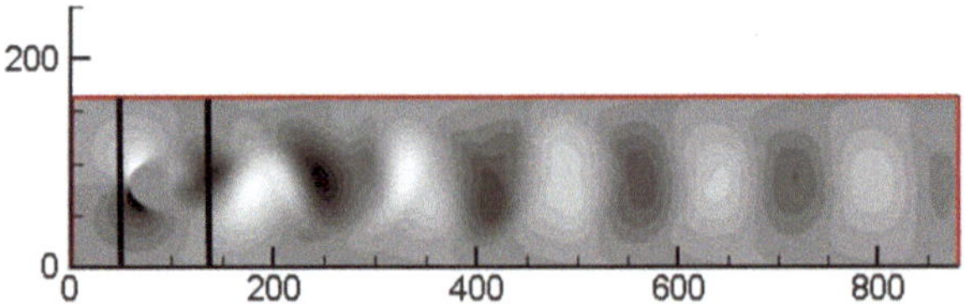

Fig. 3.7 Vertical velocity contours, and location of the fine block with respect to the cylinder at coarse time step 3.6×10^4

coarse time steps 3.7×10^4 and 4.0×10^4. The lift and the drag coefficients were calculated using the following formulae, respectively (Schafer and Turek 1996):

$$C_L = \frac{2F_L}{\rho \overline{U}^2 D}$$
$$C_D = \frac{2F_D}{\rho \overline{U}^2 D} \tag{3.6}$$

The lift and the drag forces were computed by the following equations, respectively:

$$F_L = -\int_S (\mu \frac{\partial v_t}{\partial n} n_x + P n_y) dS$$
$$F_D = \int_S (\mu \frac{\partial v_t}{\partial n} n_y - P n_x) dS \tag{3.7}$$

where F_L was the lift force, F_D was the drag force, μ was the fluid dynamic viscosity, P was the local pressure, v_t was the tangential velocity, and n_x, n_y were the x and y components of the normal to the surface of the cylinder S.

The comparison in Fig. 3.6 shows a reasonable agreement between the two cases and the lift coefficients were not symmetrical with respect to the x axis. This was due to the asymmetrical placement of the cylinder (Yu et al. 2002). The maximum absolute values for the lift coefficient, which was in the negative region, were different in both simulations (0.98 for the fixed block, and 1.03 for the moving block); however both maximum values agreed well with the values given by Schafer and Turek (1996) (0.99–1.01). The drag coefficients were $3.03 \leq C_D \leq 3.14$ for the fixed block and $3.02 \leq C_D \leq 3.138$ for the migrating block. Both maximum values were little below those reported by Schafer and Turek (1996) ($C_{D,\max} = 3.22 - 3.24$). Figure 3.7 (see Movie C3.1.2.1) shows the vertical velocity contour and the fine block position after the seventh block shift at coarse time step 3.6×10^4.

To test the quality of the data transfer through the grid interfaces and the effects of the fine block migration on the model results, the mass flux and the momentum flux were calculated at the grid interface downstream of the cylinder as shown in Fig. 3.8. The data were collected from the overlapping coarse and fine nodes of the

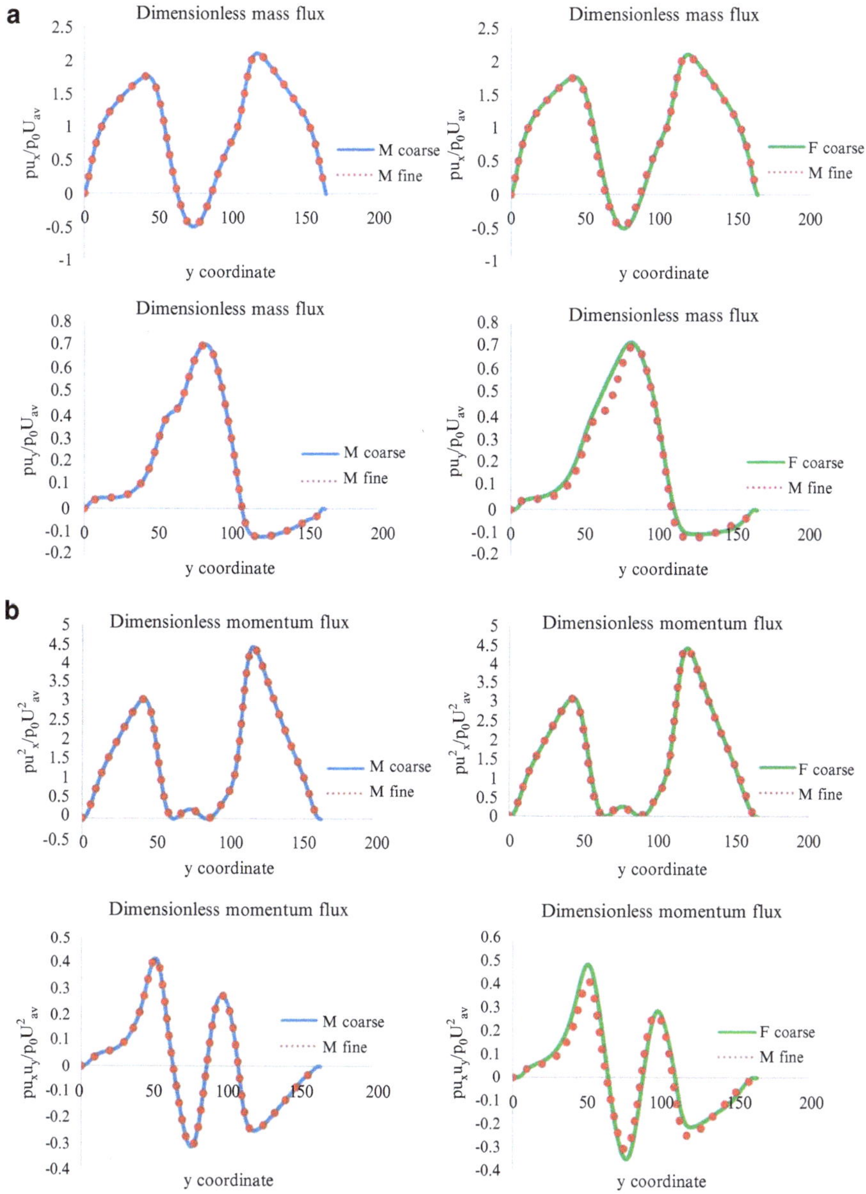

Fig. 3.8 (**a**) Graphs for the dimensionless mass flux at coarse time step 3.5×10^4, calculated for checking the quality of the data transfer through the grid interface between the fine bock and the downstream coarse block in the migrating multiblock model. Comparison between the results of the moving fine grid interface's nodes with those collected from the fixed multiblock coarse nodes which occupy the same locations. *M* and *F* in the figure stand for moving and fixed blocks, respectively. (**b**) Graphs for the dimensionless momentum flux at coarse time step 3.5×10^4, calculated for checking the quality of the data transfer through the grid interface between the fine bock and the downstream coarse block in the migrating multiblock model. Comparison between the results of the moving fine grid interface's nodes with those collected from the fixed multiblock coarse nodes which occupy the same locations. *M* and *F* in the figure stand for moving and fixed blocks, respectively

migrating block at the grid interface, and from the fixed multiblock nodes which occupied the same spatial locations. The good match between the results of the fine and the coarse block calculations at the migrating block interface was a measure of the used interpolation scheme's accuracy, and it showed that the migrating block algorithm did not alter the outcome of the calculations. The slight difference in the y components between the migrating and the fixed block results was due to the difference in the grid size, used for calculating the fluxes in both cases, and also due to the fact that the nodes where the data collection occurred in the fixed multiblock were not grid intersection nodes.

3.1.2.2 Lift of a Neutrally Buoyant Drop in Parabolic Flow

The study of multiphase flows at low to moderate Reynolds and Weber numbers, where the effect of gravity is neglected, is of interest in many applications such as the study of drop suspension in microgravity and in the study of microfluidics. In very low gravity shear and parabolic flows, the hydrodynamic lift force becomes very important, since it is no longer overshadowed by the buoyancy. The lift force is due to the hydrodynamic interaction of the drop with neighboring boundary (Halliday et al. 2005) or it is caused by a secondary velocity field at the drop surface (Legendre and Magnaudet 1998).

The goal of this section's simulation was to validate the presented model by comparing the results for the lateral migration of a 2D neutrally buoyant drop placed near a wall in parabolic flow with other numerical work. The other goal was to assess the quality of the presented migrating multiblock model results for the lift trajectory and velocity, in comparison with those from the standard Gunstensen model. The approach for the estimation of the shear lift velocity was based on tracking the mass center of the drop. The result was a displacement–time function used for the calculation of the drop lateral velocity. The quality of the measurements depended heavily on the nature of the grid, since the lift force was small likewise the change in the lateral position of the drop mass center. To minimize the effect of periodicity in the flow direction while attaining the drop equilibrium distance from the wall, a longer channel was required. For a better interface representation it was crucial to refine the grid surrounding the drop. All of this resulted in a high computational cost for the standard LBM; meanwhile it provided a good test ground for the presented migrating multiblock method.

Mortazavi and Tryggvason (2000) carried out a thorough numerical investigation of the drop shear lift in Poiseuille flow. For the case in which the ratio of the drop radius to the channel height was given as $\zeta = 0.125$, the viscosity ratio $\eta = \frac{\mu_d}{\mu_s} = 8$, the drop Reynolds number $\mathrm{Re}_d = 10.0$, and the Weber number $We = 16$, a normalized equilibrium distance from the wall of $\frac{y_{eq}}{H} \approx 0.30$ was reported.

The drop behavior in parabolic flows is characterized by the following dimensionless numbers. The channel Reynolds number:

$$\mathrm{Re}_{ch} = \frac{\overline{U}H}{\nu} \tag{3.8}$$

where H is the channel height and $\overline{U}$ is the flow average velocity.

The drop Reynolds number is given by:

$$\mathrm{Re}_{d} = \frac{\overline{U}d}{\nu} \tag{3.9}$$

The Weber number is expressed as follows:

$$We = \frac{\rho\overline{U}^2 d}{\alpha} \tag{3.10}$$

A domain made of 287,400 coarse lattice nodes and 168 by 300 fine lattice nodes, covered a drop with diameter $d = 76$ fine lattice units, yielding a ratio $\zeta = d/2H \approx 0.125$. The drop was placed at coordinate (94, 245) measured in fine lattice nodes, the density of both fluids was set to $\rho = 0.514$ and the surface tension parameter to $\alpha = 1.0 \times 10^{-4}$. The relaxation times for the ambient fluid in the fine and coarse grids were set to $\tau_f = 0.646$ and $\tau_c = 0.573$, respectively. The drop relaxation time was $\tau_d = 1.666$ leading to a viscosity ratio $\eta = 8$. The grid ratio between the coarse and fine blocks was $m = 2$. A constant force $|\boldsymbol{F}| = 2.14 \times 10^{-8}$ was used in Eq. (2.20) to induce a flow with an average velocity $\overline{U} = 0.0064$, a drop Reynolds number $\mathrm{Re}_d = 10.0$, and Weber number $We \approx 16$. The bounce back condition was applied on the upper and the lower walls, and the periodic condition was imposed at the inlet and the outlet boundaries. In the migrating multiblock the following equalities were required in the upstream coarse block after streaming:

$$\begin{aligned} f_i^{c,u}(x_{first}, y, 1) &= f_i^{c,d}(x_{last}, y, 1) \\ f_i^{c,u}(x_{first}, y, 2) &= f_i^{c,d}(x_{last}, y, 2) \\ f_i^{c,u}(x_{first}, y, 8) &= f_i^{c,d}(x_{last}, y, 8) \end{aligned} \tag{3.11}$$

where $f^{c,u}$ and $f^{c,d}$ are the distribution functions in the upstream and downstream blocks, respectively. x_{first} and x_{last} refer to the first and the last fluid nodes in the horizontal direction; the numbers indicate the lattice directions. In the downstream coarse block the following was applied:

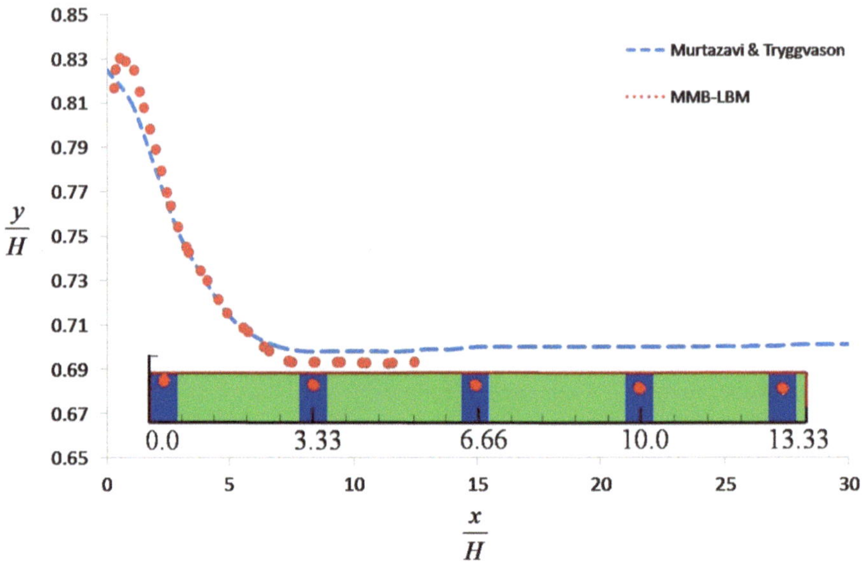

Fig. 3.9 Drop mass center normalized lateral displacement by the migrating multiblock LBM compared with the solution of Mortazavi and Tryggvason (2000) for the case with $\mathrm{Re}_d = 10.0$, $We = 16$, $\lambda = 8$, and $\zeta = 0.125$. No further data was provided for $x/H > 13.3$ since the drop reached the end of the domain 4,000 × 300 in the MMB model measured in fine lattice. The *inset* in the figure is for the phase field contours of the droplet with superimposed snapshot from different time steps

$$\begin{aligned} f_i^{c,d}(x_{last}, y, 4) &= f_i^{c,u}(x_{first}, y, 4) \\ f_i^{c,d}(x_{last}, y, 5) &= f_i^{c,u}(x_{first}, y, 5) \\ f_i^{c,d}(x_{last}, y, 6) &= f_i^{c,u}(x_{first}, y, 6) \end{aligned} \tag{3.12}$$

The source term of Eq. (2.20) was augmented by the grid ratio m in the coarse blocks as follows:

$$\phi_i = \omega_i \frac{m}{k_2} \boldsymbol{F} \cdot \boldsymbol{c}_i \tag{3.13}$$

The drop center of gravity normalized position with respect to the drop axial normalized position is shown in Fig. 3.9 (see Movie C3.1.2.2—MIGrho). The migrating multiblock result was compared with the solution of Mortazavi and Tryggvason (2000). The normalized equilibrium distance resulting from the migrating multiblock was $\frac{y_{eq}}{H} \approx 0.31$. The presented model results were fairly good, in comparison with those of Mortazavi and Tryggvason (2000).

To compare the results of the presented model with those of the standard LBM, the same flow condition and geometric settings were used for a domain consisting of 2,000 by 300 lattice squares to avoid excessive computations in the standard LBM.

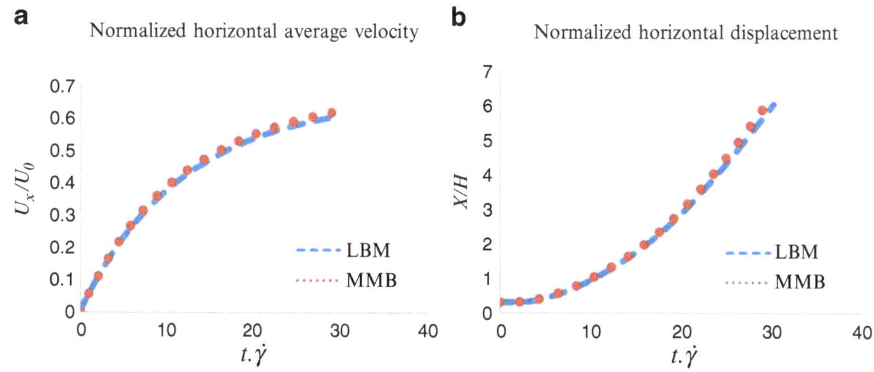

Fig. 3.10 Normalized fluid average horizontal velocities (**a**), and normalized drop mass center displacements in the flow direction (**b**), for both the migrating block and the standard Gunstensen model versus dimensionless time

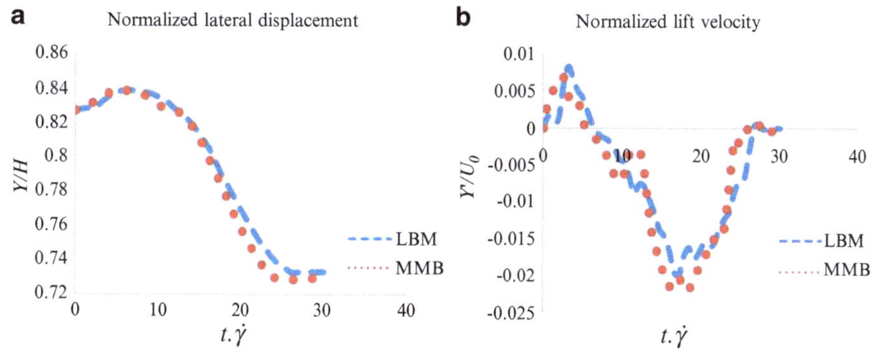

Fig. 3.11 Normalized lateral trajectory of the drop mass center (**a**) for the migrating block and the standard LBM measured with respect to dimensionless time. Normalized lift velocity (**b**) for the migrating block and the standard LBM calculated with respect to the dimensionless time

The equivalent domain for the migrating block scheme consisted of 137,400 coarse lattice nodes and 168 by 300 fine lattice nodes. The drop was placed at coordinate (94, 248) measured in fine lattice nodes.

A dimensionless approach was used for the analysis of the results. H was selected as a characteristic length, $U_0 = \frac{3}{2}\overline{U}$ the undisturbed centerline flow velocity, as characteristic velocity, and the inverse shearing strain rate $\dot{\gamma}^{-1} = \frac{H}{2U_0}$ as characteristic time. The shear rate was calculated at the vertical position $3H/4$ since this position was representative of the equilibrium point in the drop lift activity space. The normalized fluid average velocities and the normalized displacements of the drop mass center in the flow direction versus dimensionless time from the two simulations are shown in Fig. 3.10.

The lateral displacements normalized by the channel width versus the dimensionless time from both simulations are shown in Fig. 3.11.The lift velocities were

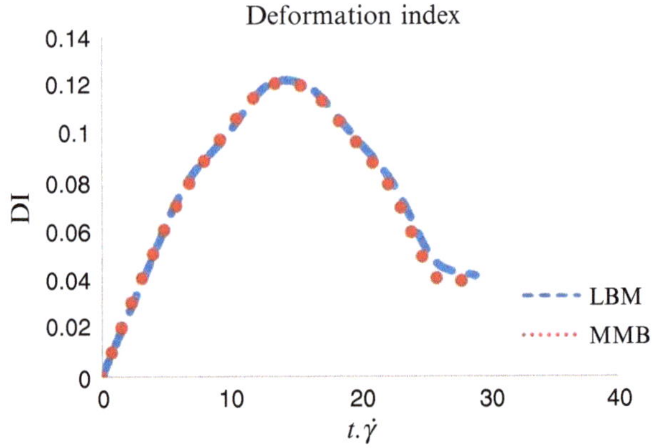

Fig. 3.12 Deformation indexes of the drops from both the standard LBM and the migrating multiblock calculated with respect to the dimensionless time

calculated from the lift displacement–time data as time derivative by a finite difference scheme with second-order accuracy, and were normalized by the centerline flow velocity as shown in Fig. 3.11.

It was clear from Fig. 3.11 that the velocities observed at $t\dot{\gamma} < 7$ should be neglected due to the drop tilt during the initialization of the simulations, and that the drop's lift velocity is an order of magnitude smaller than its translational velocity. The same was extracted in the work of Sukumaran and Siefert (2001) who studied the lift of the near-wall neutrally buoyant vesicles in shear flow. The distance from the top wall at dimensionless time step $t\dot{\gamma} = 28.8$ for the standard LBM was $\frac{y}{H} = 0.2733$ versus $\frac{y}{H} = 0.2728$ for the migrating block.

The deformation index $DI = \frac{(a-b)}{(a+b)}$, where a is the drop major axis and b is the drop minor axis, varied between the values $0 \leq DI \leq 0.12$ during the simulations as shown in Fig. 3.12. The reduction in the DI associated with time was due to the reduction in the viscous stress, when the drop drifted away from the wall leading to a reduced DI. Cox (1969) proposed a theoretical formula for the calculation of the drop deformation in a general time-dependent fluid flow with a range of capillary numbers and viscosity ratios. The time dependence of the DI was through a decaying exponential function which led after long time (steady state) to the following relationship:

$$DI = \frac{5(19\lambda + 16)}{4(\lambda + 1)\sqrt{\left(\frac{20}{Ca}\right)^2 + (19\lambda)^2}} \tag{3.14}$$

where $Ca = \mu_s \dot{\gamma} d/2\alpha$ is the capillary number. The strain rate used in the calculation of this work's capillary number was locally defined by $\dot{\gamma}(y) = \left|\frac{8U_0}{H^2}\left(\frac{H}{2} - y\right)\right|$. The deformation index calculated by Eq. (3.14) yielded $DI = 0.151$ for the dimensionless time step $\dot{\gamma}t = 12.8$ and the calculated capillary number $Ca = 0.78$. The dimensionless mass center location was $\frac{y}{H} = 0.824$, which corresponded to the location where the simulation results led to the highest value $DI = 0.12$ as

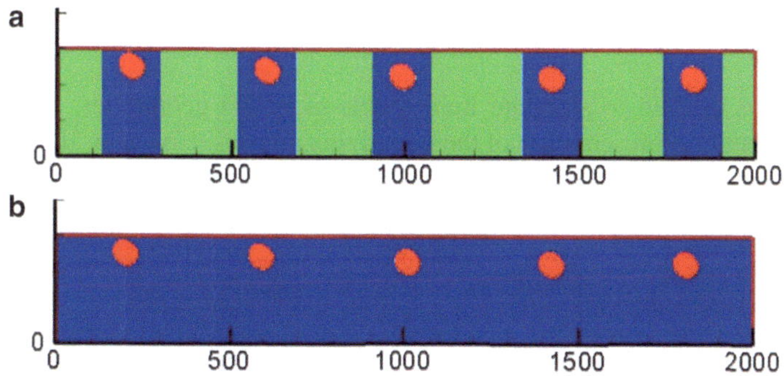

Fig. 3.13 Phase field contour for five consecutive snapshots taken at different time steps and superimposed in the figure. The *blue* blocks are fine, and the *green* blocks are coarse: (**a**) migrating multiblock and (**b**) standard LBM

shown in Fig. 3.12. The difference between the measured and the calculated deformation indices could be resulting from the transient nature of the drop deformation under the shear lift as measured from the simulation, compared to the steady-state deformation described by Eq. (3.14).

Figure 3.13 presents the phase field contours, with the various positions of the drop generated by super-imposing consecutive snapshots taken at different time steps.

To analyze the computational time advantage of the presented model, the following formula was introduced:

$$Ga = m\frac{N_x N_y}{mL_x N_y + \frac{1}{m^2}(N_x - L_x)N_y} \tag{3.15}$$

where N_x and N_y are the domain length and width measured in fine grids spacing respectively, and L_x is the length of the fine block. Equation (3.15) was based on the idea that for the calculation of one time step in the coarse blocks expressed in fine lattice units as $(N_x - L_x)N_y/m^2$, there is a need for m time steps in the fine block having dimensions $L_x N_y$ and in the standard LBM with dimensions $N_x N_y$, respectively. Equation (3.15) is applicable only for 2D models, with the fine block covering the entire width. For the current simulation the formula leads to a time gain $Ga = 5.04$. This was also confirmed by comparing the computational time required for the simulations using the standard Gunstensen model and the presented migrating multiblock method simultaneously. Using DELL Precision 490 workstation, one time step in the standard model required 0.516 second for execution, while the time needed for the same time step in the migrating multiblock was 0.108 second leading to $Ga = \frac{t_{st}}{t_{mmb}} = 4.77$. The difference between the calculated and the measured computational time gain could be used to evaluate the code level of efficiency.

3.2 Migrating Multiblock Scheme for the D3Q19 LBM

It is always desired to simulate fluid problems in 3D geometries, because they provide a better representation of the underlying physics. However 3D simulations are computationally costly. This prompted the extension of the migrating multiblock to 3D geometries.

3.2.1 *Modified Model with Density Contrast*

The standard Gunstensen model does not tolerate a density contrast between the constituent fluids of the mixture; meanwhile in this section the density contrast is required for the simulation of a rising bubble in infinite medium. Therefore the method of Grunau et al. (1993) is used with some suitable modifications for this purpose. The main collision step is described by two distribution functions instead of the standard Gunstensen color blind function $f_i(\boldsymbol{x}, t)$, and it is expressed through the following:

$$\widehat{f}_i^k(\boldsymbol{x}, t+\delta_t) = f_i^k(\boldsymbol{x}, t) - \frac{1}{\tau^k}\left[f_i^k(\boldsymbol{x}, t) - f_i^{k,eq}(\rho, \rho\boldsymbol{u})\right] + \phi_i(\boldsymbol{x}) \tag{3.16}$$

where $\widehat{f}_i^k$ refers to postcollision distribution functions for the lighter and heavier fluids ($k = L, H$), τ^k are the relaxation times for both fluids, and $\phi_i(x)$ is a source term used for the introduction of a force into the fluid, which could be utilized to induce the necessary surface tension and to account for the difference in density between the various fluids, thus creating buoyancy force. The surface tension is created by the method of Lishchuk et al. (2003) which imposes a normal force at the interface by Eq. (2.16). The choice of the equilibrium distribution functions is in principle arbitrary provided that these functions satisfy the following mass and momentum conservation laws:

$$\begin{aligned} \rho^L(\boldsymbol{x}, t) &= \sum_{i=0}^{Q-1} f_i^L = \sum_{i=0}^{Q-1} f_i^{L,eq} \\ \rho^H(\boldsymbol{x}, t) &= \sum_{i=0}^{Q-1} f_i^H = \sum_{i=0}^{Q-1} f_i^{H,eq} \\ \rho\boldsymbol{u} &= \sum_{i,k} \boldsymbol{c}_i f_i^k = \sum_{i,k} \boldsymbol{c}_i f_i^{k,eq} \end{aligned} \tag{3.17}$$

where $\rho = \rho^L + \rho^H$ is the total density, ρ^L is density of the light fluid, and ρ^H is the density of the heavier fluid. The following equilibrium distribution functions are used in the presented model and they were taken from the work of Maini (2007) due to the simplicity of their derivation and application:

$$
\begin{aligned}
f_i^{L,eq} &= \rho^L \omega_i \left[1 + \frac{3}{c^2} \boldsymbol{c}_i.\boldsymbol{u} + \frac{9}{2c^4} (\boldsymbol{c}_i.\boldsymbol{u})^2 - \frac{3}{2c^2} \boldsymbol{u}.\boldsymbol{u}\right] \\
f_i^{H,eq} &= \rho^H \omega_i \left[r_i + \frac{3}{c^2} \boldsymbol{c}_i.\boldsymbol{u} + \frac{9}{2c^4} (\boldsymbol{c}_i.\boldsymbol{u})^2 - \frac{3}{2c^2} \boldsymbol{u}.\boldsymbol{u}\right]
\end{aligned}
\tag{3.18}
$$

where the constant is calculated as follows:

$$
r_i = \left\{ \begin{array}{l} 3 - 2\gamma, \rightarrow i = 0 \\ \gamma, \rightarrow i \neq 0 \end{array} \right\}
\tag{3.19}
$$

$\gamma = \frac{\rho^L}{\rho^H} = \frac{(c_s^H)^2}{(c_s^L)^2}$ is the dimensionless parameter for the density ratio. c_s^L, c_s^H are the speeds of sound used in conjunction with τ^k for determining the kinematic viscosities of the two fluids. This is realized through the use of Eq. (2.5). The relationship between the source term $\phi_i(\boldsymbol{x})$ and a constant macroscopic force $\boldsymbol{F}(\boldsymbol{x})$ such as a body force is expressed by Eq. (2.20). The ratio between the various fluids weighting constants is expressed through the following relationship $\omega_0^H = \omega_0^L(3 - 2\gamma)$; $\omega_{i\neq 0}^H = \gamma \omega_{i\neq 0}^L$.

In the presented model, the postcollision distribution functions are calculated by Eq. (3.16) for each fluid using a sweep throughout the whole domain. This will not alter by any means the light and heavy fluid mix at the interface and it allows the use of the appropriate equilibrium distribution function, based on the right proportion of masses present at the individual interfacial nodes from both fluids. The total color blind postcollision distribution function is hence after invoked using $\widehat{f}_i = \widehat{f}_i^L + \widehat{f}_i^H$. This paves the way for the segregation process which is achieved by not allowing color diffusion at the fluid interfacial nodes. The formulaic method of D'Ortona et al. (1995) is applied in the model using Halliday et al. (2007) generalized formula for the implementation of the method in 3D models:

$$
\begin{aligned}
\widehat{\widehat{f}}_i^L (\boldsymbol{x}, t + \delta_t) &= \frac{\rho^L}{\rho^L + \rho^H} \widehat{f}_i(\boldsymbol{x}, t + \delta_t) + \beta \frac{\rho^L \rho^H}{\rho^L + \rho^H} \omega_i \nabla \rho^N \cdot \boldsymbol{c}_i \\
\widehat{\widehat{f}}_i^H (\boldsymbol{x}, t + \delta_t) &= \widehat{f}_i(\boldsymbol{x}, t + \delta_t) - \widehat{\widehat{f}}_i^L (\boldsymbol{x}, t + \delta_t)
\end{aligned}
\tag{3.20}
$$

β is the segregation parameter. $\widehat{\widehat{f}}_i^L$, $\widehat{\widehat{f}}_i^H$ refer to postcollision postsegregation distribution functions of the light (red) and heavy (blue) fluids respectively. The streaming step follows the segregation step through the following:

$$
\begin{aligned}
f_i^L(\boldsymbol{x} + \boldsymbol{c}_i\delta_t, t + \delta_t) &= \widehat{\widehat{f}}_i^L (\boldsymbol{x}, t + \delta_t) \\
f_i^H(\boldsymbol{x} + \boldsymbol{c}_i\delta_t, t + \delta_t) &= \widehat{\widehat{f}}_i^H (\boldsymbol{x}, t + \delta_t)
\end{aligned}
\tag{3.21}
$$

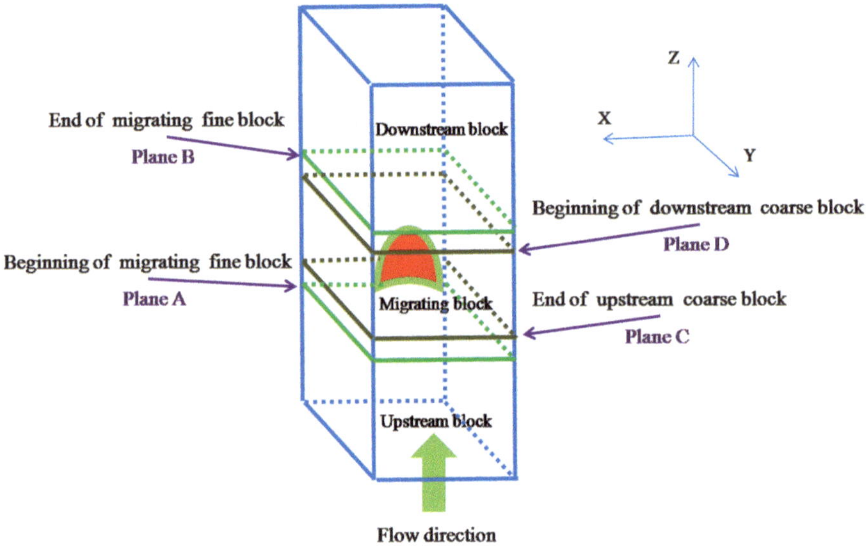

Fig. 3.14 Illustration of the 3D multiblock lattice Boltzmann method domain comprising two coarse blocks and one fine block

3.2.2 The Migrating Multiblock Algorithm

A 3D domain shown in Fig. 3.14 describes three blocks in which the central is cast with fine grid and the others are with coarse grid. The spatial and temporal ratio of the lattice spacing for both fine and coarse grid m is defined by Eq. (2.39).

The relaxation times are linked through Eq. (2.40) which guarantees uniform viscosity throughout the fluid in all blocks (Yu et al. 2002). Each grid interface is formed by two sets of overlapping planes of coarse and fine nodes (A, C and B, D) and one additional plane of fine nodes in between for each interface (not shown in Fig. 3.14).

The transfer of information between the various grids is executed on all interfacial planes by the same formulae which were implemented in the 2D model and were extracted from the work of Yu et al. (2002). At the fine plane (A and B) Eq. (2.41) is used for the transfer. At the coarse plane (C and D) the transfer is executed by Eq. (2.42).

Symmetric cubic spline interpolation is needed for calculating $\widehat{f}_i^f$ on those fine nodes designated with asterisks and red circles in Fig. 3.15 by Eqs. (3.1) and (3.2). Spatial and temporal interpolations are only needed at planes (A and B) illustrated in Fig. 3.15. A constructive suggestion on the sequence used for the spatial interpolation is to start first by calculating the values of the postcollision distribution functions at the asterisks positions following the y direction through a sweep by x coarse spatial steps. This is followed by the calculation of the postcollision distribution functions at the red circles in the x direction through a sweep by y fine spatial steps.

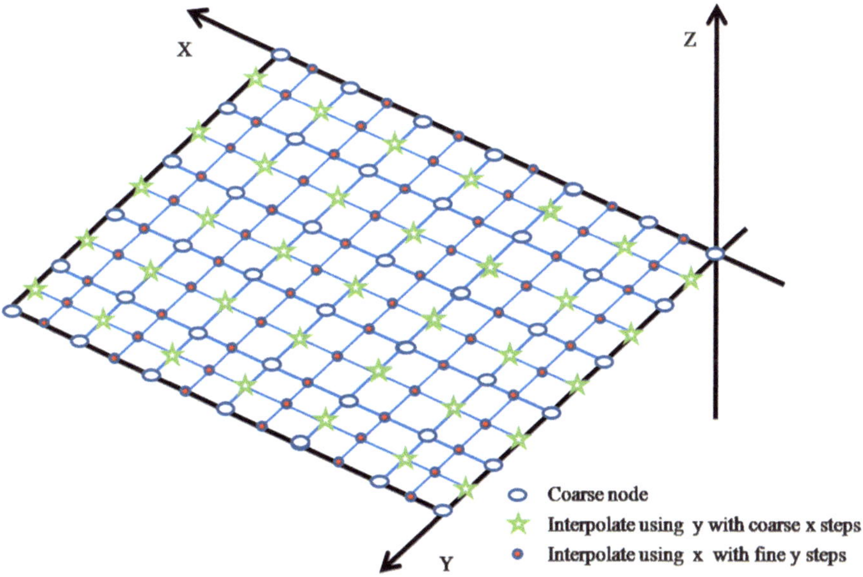

Fig. 3.15 Illustration of the grid interface plane at the beginning or the end of the fine block, where spatial and temporal interpolation are required

A three-point Lagrangian temporal interpolation is required for all fine nodes shown in Fig. 3.15 in order to synchronize the solution and it is executed by Eq. (2.43).

The migrating multiblock algorithm is based on the idea of exchanging nodes type at the grid interfaces. This is driven by the necessity of covering the fluid interface with fine grid throughout the simulation time. The condition for the occurrence of the node type exchange is implied by the displacement of the interface in the flow direction by a value greater than one coarse unit. The exchange requires the creation of two planes of fine nodes and the extinction of one plane of coarse node in the downstream block, and the creation of one plane of coarse nodes and diminishing two planes of fine nodes in the upstream block. The creation of new nodes requires the use of extrapolation method which could be of first or second order. The aim is to minimize numerical diffusion, which is usually marginal, since the newly created nodes undergo information transfer by Eqs. (2.41) and (2.42), before proceeding with any further calculation.

3.2.3 Simulation Results and Discussion

The velocity and the shape of a light bubble moving in a denser fluid are strongly dependent on the ratio among three forces—the driving force from buoyancy, the resistance from the viscous friction, and the surface tension which tries to maintain the shape of the bubble spherical. For the simulation of a free rising bubble in infinite medium with periodic boundary condition the effects from the noted forces can be represented by the following dimensionless parameters (Tolke et al. 2002):

$$E_o = \frac{g\Delta\rho d^2}{\sigma_0}, M_o = \frac{g\mu^4\Delta\rho}{(\rho^H)^2{\sigma_0}^3}, \mathrm{Re} = \frac{U_T d\rho^H}{\mu} \tag{3.22}$$

Here g is the acceleration due to gravity, d is the bubble diameter, ρ^H is the density of the heavy fluid, and σ_0 is the surface tension at the interface. The Eotvos number E_o is the ratio between buoyancy and the surface tension, the Morton number M_o compares the inertial effects, viscous drag and the surface tension, the Reynolds number Re represents the ratio between the inertial force and the viscous drag. The interplay between these parameters determines the shape of the bubble which could vary from spherical, ellipsoidal, ellipsoidal cap, spherical cap, and eventually to skirted as this was demonstrated through the experimental work of Bhaga and Weber (1981).

3.2.3.1 Terminal Velocity

The terminal velocity of a rising bubble in infinite medium with $E_o > 40$ and $M_o > 200$ can be estimated analytically through solving the following equation with respect to the Reynolds number (Clift et al. 2005):

$$2\mathrm{Re}^2 + 6\mathrm{Re}\frac{2+3\eta}{1+\eta} - E_o^{3/2}M_o^{-1/2} = 0 \tag{3.23}$$

where $\eta = \mu_L/\mu_H$ is the viscosity ratio for the light and heavier fluids.

The demonstration of the presented method is performed as follows: The simulation domain was 61×61×351 measured in fine lattice units, and the bubble initial radius was $d = 12$ lattice fine units as shown in Fig. 3.16 (see Movies C3.2.3.1—MBouy1 and C3.2.3.1—MBouy2).

Periodic condition was imposed in all directions. The fine block consisted of 61×61×48 lattice cubes yielding 178,608 fine nodes; meanwhile the upstream and the downstream blocks comprised of 140,933 coarse nodes. The spacing ratio was $m = 2$, the relaxation times were $\tau_f = 0.9$ and $\tau_c = 0.7$ for the fine and coarse blocks respectively, and the segregation parameter was $\beta = 0.55$. A density $\rho^H = 2$, density ratio $\gamma = 0.5$ leading to a density contrast $\gamma^{-1} = 2.0$, kinematic viscosity ratio $\eta = 1$, dynamic viscosity $\mu = 0.266$, and a surface tension $\alpha = 0.001$ were used in five cases where the gravitational acceleration g was varied consecutively.

The values for the acceleration g from five different cases and their corresponding Eotvos, Morton, Reynolds numbers, and theoretical terminal velocities are presented in Table 3.1.

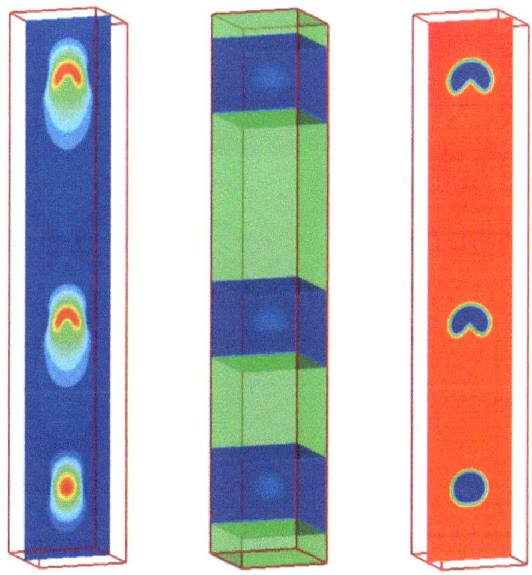

Fig. 3.16 3D simulation of a rising bubble in an infinite medium with $M_0 = 821$, $E_0 = 374$, and Re $= 8.10$, velocity contour in the z direction for superimposed snapshots at three different time steps (*left*), 3D phase field contours where the fine migrating superimposed blocks are shown in *blue* and the coarse blocks in *green* (*center*), 2D cut of the density contour depicting the droplet profile during the various time steps (*right*)

Table 3.1 Simulation results for five different cases, U_T terminal velocity from Eq. (3.23), U_M terminal velocity from the numerical simulation

g	E_0	M_0	Re	U_T	U_M
0.00036	207	455	5.42	0.0301	0.0304
0.00045	259	568	6.33	0.0351	0.0359
0.00054	311	682	7.16	0.0397	0.0401
0.00065	374	821	8.10	0.0450	0.0444
0.00078	449	985	9.125	0.0506	0.0489

A comparison between the terminal velocities calculated by the semianalytical Eq. (3.23) and the model terminal velocities for the various cases is shown in Fig. 3.17.

The figure also shows the effects of the change in the dimensionless parameters of Eq. (3.22) and in particular the Eotvos number due to the change in the gravitational force on the steady-state shape of the rising bubble.

3.2.3.2 Bubble Shape

Grace (1973) brought together the results of several experiments which focused on the case of a single rising bubble in infinite media.

Grace delineated three major regions where the shape of the bubble was determined by the dimensionless parameters of Eq. (3.22). The regions were as follows: spherical bubble, ellipsoidal bubble, and spherical cap regions. The various regions were demarcated in a shape regime map shown in Fig. 3.18. This map was further expanded by Bhaga and Weber (1991) and more distinct regions were added which included oblate ellipsoidal, oblate ellipsoidal cap, and skirted bubble shapes.

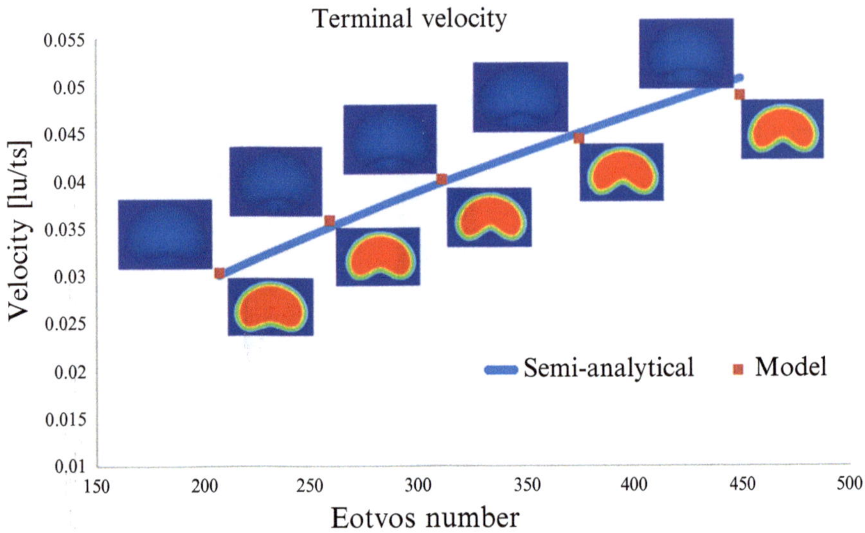

Fig. 3.17 Terminal velocity comparison and shape change with respect to the Eotvos number for values stated in Table 3.1

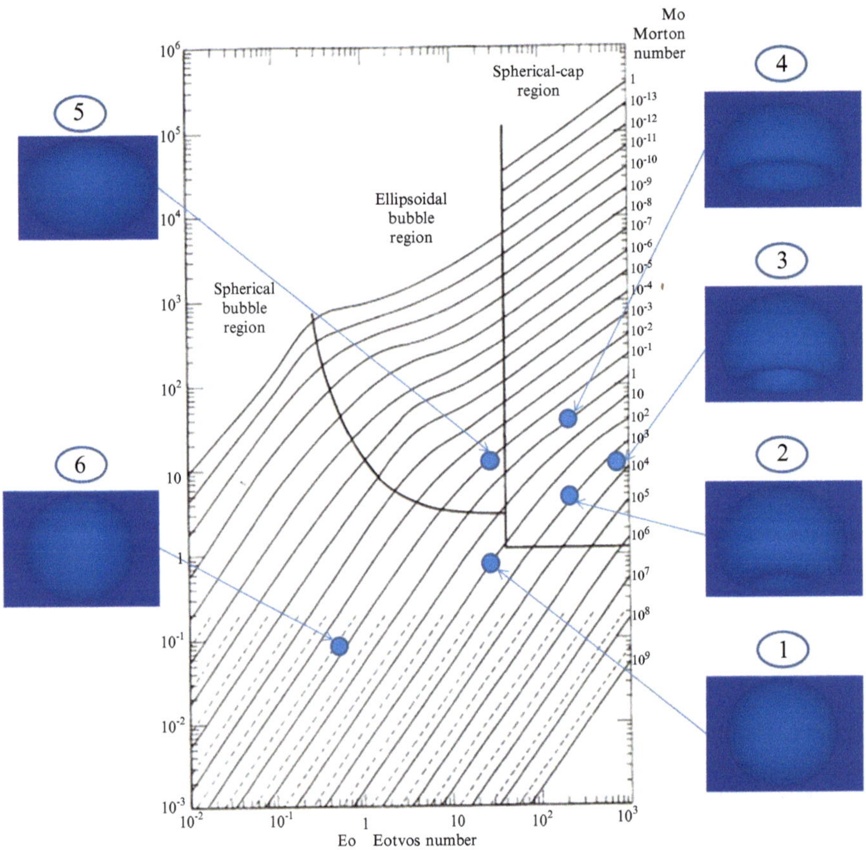

Fig. 3.18 Shape regime map by Grace (1973) used for locating the presented model results for the various cases presented in Table 3.2. The model shows good fit within the three shape map regions

Table 3.2 Variables and dimensionless numbers for a few shape region simulations

g	E_O	M_O	Re	μ	Region
0.00005	29	63	0.63	0.2666	SBR-1
0.00036	207	455	5.42	0.2666	SCR-2
0.00015	864	1,894	11.70	0.2666	SCR-3
0.00036	207	0.73	33.32	0.0533	SCR-4
0.00005	29	0.10	12.60	0.0533	EBR-5

SBR spherical bubble region, *SCR* spherical cap region, *EBR* ellipsoidal bubble region

Data from six simulations are presented in Table 3.2. The shapes obtained by the presented model for the parameters of Table 3.2 were plotted with their corresponding location in the three map regions. The results as seen from Fig. 3.18 indicate that the model yielded droplet shapes which matched well the experimental observation at the various regions of Grace's shape regime map.

3.2.3.3 Density Ratio

Although the presented algorithm deviated slightly from the original Gunstensen model in order to tolerate a density contrast among the constituent fluids, the density contrast which was tried here was $\gamma^{-1} = 10$. The main concern emanated from the formation of a thick interface in the phase field presentation of the suspended phase. A density ratio $\gamma = 0.1$, segregation parameter $\beta = 0.15$, and a gravitational constant $g = 8.5 \times 10^{-5}$ were used in a simulation, in which the domain and all other fluid properties were similar to those used in the terminal velocity section. This led to Re $= 4.2$, $M_o = 193.23$, and $E_o = 88.13$. The outcome of the analytical formula for the terminal velocity differed from the simulation result since the formula was not applicable for $M_o < 200$; however the bubble shape and despite the thick interface still fitted well Grace's shape regime map and it fell in the spherical cap region.

Figure 3.19 shows the vertical velocity, the phase field, and the density contours of a rising bubble with a density contrast $\gamma^{-1} = 10$.

To reduce the thickness of the interface and to allow a higher density ratio, the formula for segregation Eq. (3.20) was modified to match the original formula of D'Ortona et al. (1995):

$$
\begin{aligned}
\widehat{\widehat{f}}_i^L(\boldsymbol{x}, t+\delta_t) &= \frac{\rho^L}{\rho^L+\rho^H}\widehat{f}_i(\boldsymbol{x}, t+\delta_t) + \beta\frac{\rho^L\rho^H}{(\rho^L+\rho^H)^2}\omega_i\nabla\rho^N\cdot\boldsymbol{c}_i \\
\widehat{\widehat{f}}_i^H(\boldsymbol{x}, t+\delta_t) &= \widehat{f}_i(\boldsymbol{x}, t+\delta_t) - \widehat{\widehat{f}}_i^L(\boldsymbol{x}, t+\delta_t)
\end{aligned}
\tag{3.24}
$$

With the right selection of the values for the density ratio γ and the segregation parameter β, higher density ratios are achievable with a good interface thickness.

The vertical velocity, phase field, and the density contours of a buoyant droplet are shown in Fig. 3.20. A density ratio $\gamma^{-1} = 20$, segregation parameter $\beta = 0.1$,

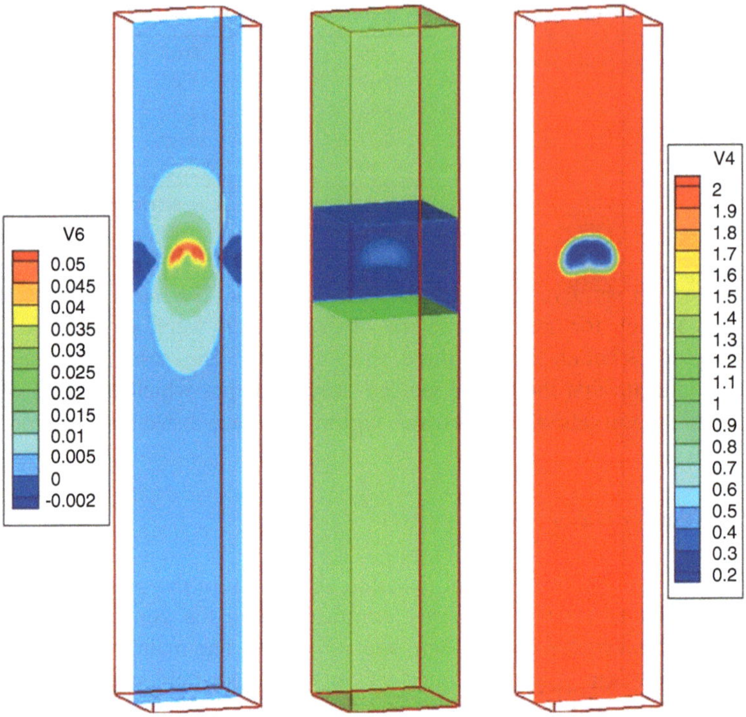

Fig. 3.19 Vertical velocity contour (*left*), phase field contour (*center*), and density contour (*right*) for a rising bubble with a density contrast $\gamma^{-1} = 10$, $M_0 = 193$, $E_0 = 88$, and Re $= 4.2$

and a gravitational constant $g = 8.5 \times 10^{-5}$ were used in this simulation. It is clear from Fig. 3.20 that the interface of the droplet is well controlled due to the application of Eq. (3.24) for segregating the fluid.

3.2.3.4 Multiple Bubbles

To illustrate the migrating multiblock capability of dealing with multiple bubbles seeded in the fine block, a two trailing bubbles case was studied. The two bubbles were initialized in the same domain described by the fifth case from Table 3.1, with the bubbles mass centers lying on the domain central vertical axis and separated by a distance of $2d$. The fine block was made of $61 \times 61 \times$fine lattice units, to accommodate for the two bubbles.

The upper bubble moved in a quiescent liquid while the lower bubble followed into a low-drag region resulting from the wake caused by the upper bubble. Therefore the two bubbles had different deformation characteristics, and a nonzero relative velocity diminished the separating distance between them, which

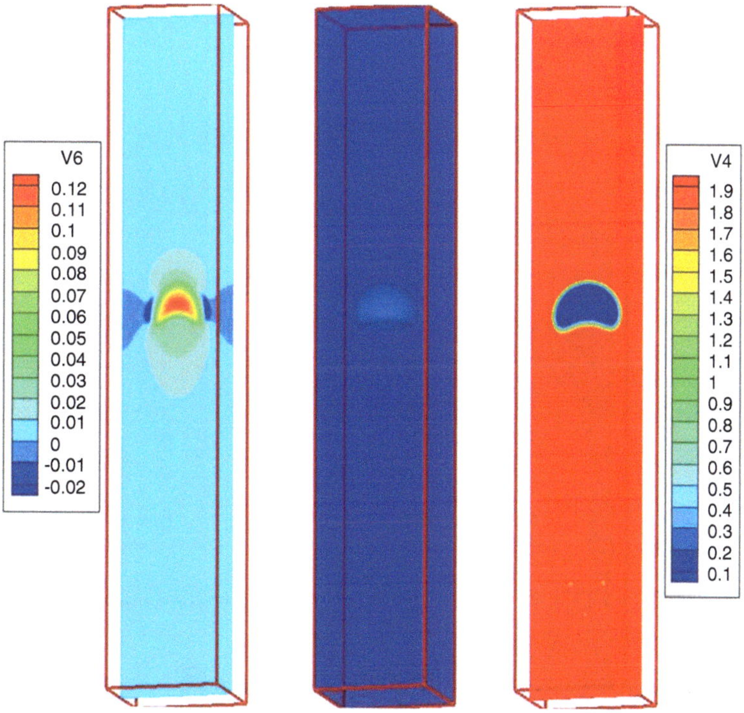

Fig. 3.20 Vertical velocity contour (*left*), phase field contour (*center*), and density contour (*right*) for a rising bubble with a density contrast $\gamma^{-1} = 20$, $M_0 = 193$, $E_0 = 88$, and Re $= 4.2$

eventually led to their collision and coalescence. Figure 3.21 shows the vertical velocity contour of the trailing bubbles which indicates a higher velocity of the lower bubble in comparison with the velocity of the upper bubble. Figure 3.21 shows also the phase field contour and the density contour of the trailing bubbles.

The snapshots in Fig. 3.22 for the phase field contours taken at different time steps suggest that the coalescence of the two bubbles did not occur instantaneously and that some liquid was trapped inside the newly formed bubble due to the high inertia impact and to the lack of enough time for the liquid to squeeze out of the bubble (see Movie 3.2.3.4—Trailing). This was also observed by Gupta and Kumar (2008). The liquid eventually drained off the bubble, hence allowing it to attain its steady size which was double the original size of the seeded droplets. Figure 3.22 shows the change in the dimensionless distance measured between the tops of the lower and the upper bubbles. The graph in Fig. 3.22 agreed qualitatively with the results of Takada et al. (2001). This simulation suggests that the presented model is capable of handling multiple bubbles clustered in the fine blocks; however it is unsuitable for solving problems with bubbles scattered randomly inside the whole domain.

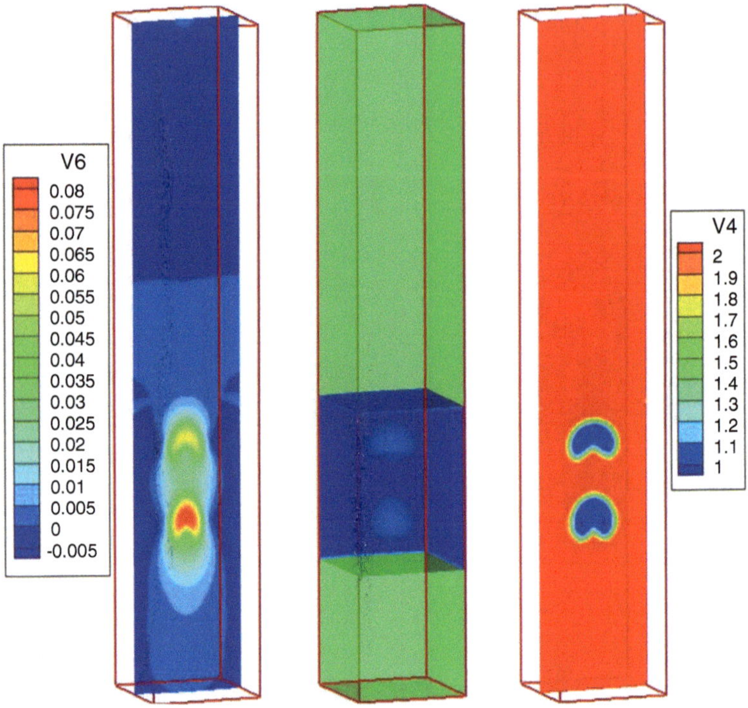

Fig. 3.21 Vertical velocity contour (*left*), phase field contour (*center*), and density contour (*right*) for two trailing bubbles with a density contrast $\gamma^{-1} = 2$, $M_0 = 985$, and $E_0 = 449$

3.2.3.5 Grid Interface Data Transfer Integrity Test

The accuracy of the presented scheme depends heavily on the quality of the data transfer required at the various grid interfaces. This transfer is directly influenced by the schemes used for both the spatial and the temporal interpolation. The test of the data transfer is also relevant for investigating the effects of the moving fine block on the total outcome of the problem. A good scheme should perform well at the grid interface of the downstream block and after numerous time steps.

The fifth case from Table 3.1 was again used for the investigation of the smoothness of the data transfer at the location indicated as (plane B) in Fig. 3.14 and at 9,000 coarse time steps. The dimensionless mass flux ${}^{\rho u_z}\!/_{\rho_H U_T}$ and the dimensionless momentum flux ${}^{\rho u_z^2}\!/_{\rho_H U_T^2}$ in the vertical direction were calculated locally using the data stored at the gird interface coarse nodes, the overlapping fine nodes, and at all the fine nodes respectively. Figure 3.23 shows a qualitative comparison between the three readings for the dimensionless mass flux, while Fig. 3.24 represents a comparison between the results of the dimensionless momentum flux.

Both figures show good qualitative agreement which indicates that the migration of the fine block did not alter the quality of the data transfer at the grid interfaces.

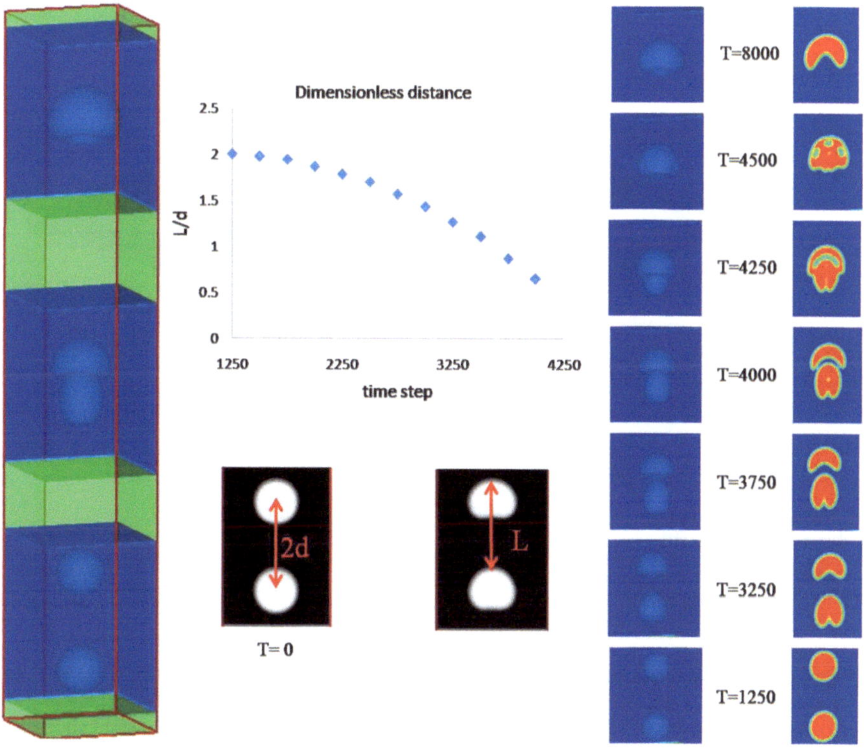

Fig. 3.22 Phase field contours of two trailing bubbles taken at different time steps and superimposed in the same frame (*left*), graph for the changing dimensionless distance between the trailing bubbles with respect to time (*center*), 3D phase field contours and their respective 2D cut views of selective snapshots intended to show various events such as trailing, collision, and coalescence of the two bubbles (*left*)

For a quantitative comparison the following composite was used to calculate both fluxes (Rostam and Mahdi 2009):

$$\iint_D f(x,y)dxdy = \frac{hk}{9}\Bigg\{ f(x_0,y_0) + 2\sum_{i=1}^{(n/2)-1} f(x_{2i},y_0) + 4\sum_{i=1}^{(n/2)} f(x_{2i-1},y_0) + f(x_n,y_0)$$

$$+2\left[\sum_{j=1}^{(m/2)-1} f\left(x_0,y_{2j}\right) + 2\sum_{j=1}^{(m/2)-1}\sum_{i=1}^{(n/2)-1} f\left(x_{2i},y_{2j}\right) + 4\sum_{j=1}^{(m/2)-1}\sum_{i=1}^{(n/2)} f\left(x_{2i-1},y_{2j}\right) + \sum_{j=1}^{(m/2)-1} f\left(x_n,y_{2j}\right)\right]$$

$$+4\left[\sum_{j=1}^{(m/2)} f\left(x_0,y_{2j-1}\right) + 2\sum_{j=1}^{(m/2)}\sum_{i=1}^{(n/2)-1} f\left(x_{2i},y_{2j-1}\right) + 4\sum_{j=1}^{(m/2)}\sum_{i=1}^{(n/2)} f\left(x_{2i-1},y_{2j-1}\right) + \sum_{j=1}^{(m/2)} f\left(x_n,y_{2j-1}\right)\right]$$

$$+f(x_0,y_m) + 2\sum_{i=1}^{(n/2)-1} f(x_{2i},y_m) + 4\sum_{i=1}^{(n/2)} f(x_{2i-1},y_m) + f(x_n,y_m)\Bigg\} \tag{3.25}$$

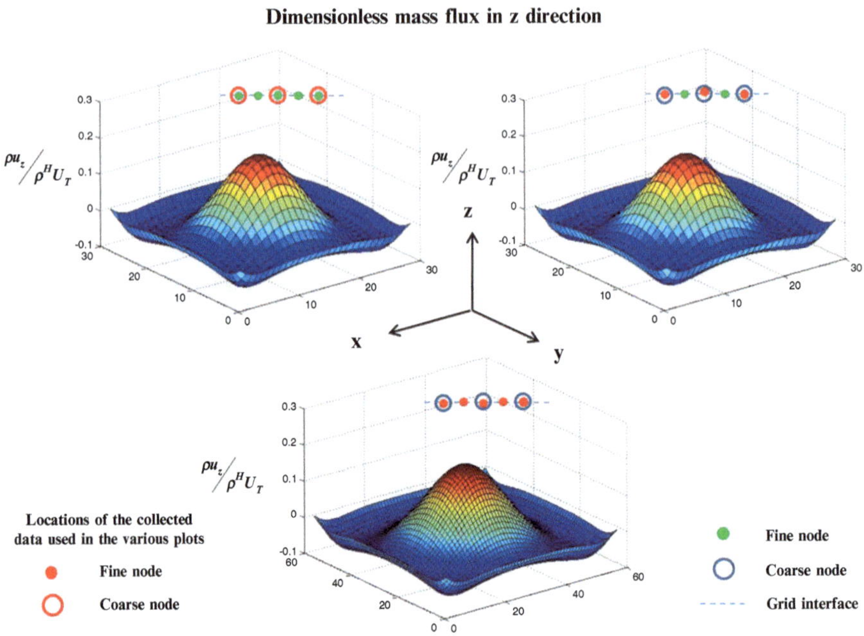

Fig. 3.23 Dimensionless mass flux measured at the upstream grid interface (plane B in Fig. 3.14) for case five from Table 3.1 at 9,000 coarse time steps; data collected from the coarse nodes (*top left*), data collected from the overlapping fine nodes (*top right*), and data collected from all the fine nodes (*bottom*)

where $h = k = 2$ lattice units and $m = n = 30$ for the coarse and the overlapping fine nodes calculations, and $h = k = 1$ lattice unit and $m = n = 60$ for the fine nodes calculation.

The results presented in Table 3.3 show a good match between the various calculations, thus indicating that the interpolation schemes used in the presented model had delivered the expected task, and that the smoothness of the data transfer was acceptable.

3.2.3.6 Computational Time Advantage

To analyze the computational time advantage of the presented model, the third case from Table 3.1 was repeated using the approach for density contrast proposed in Sect. 2.1. The grid spacing for the entire domain was similar to the spacing used for the fine block in the previous runs. The ratio between the computational times required for both cases was $Ga = t_{std}/t_{mmb} = 4.92$. The phase field contours, and the bubble vertical displacement versus time steps from the two runs are shown in Fig. 3.25.

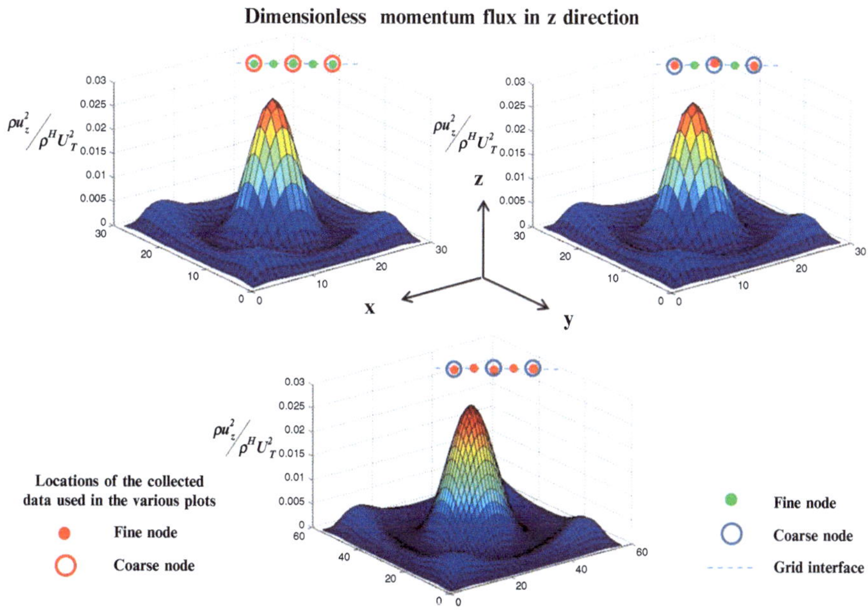

Fig. 3.24 Dimensionless momentum flux measured at the upstream grid interface (plane B in Fig. 3.14) for case five from Table 3.1 at 9,000 coarse time steps; data collected from the coarse nodes (*top left*), data collected from the overlapping fine nodes (*top right*), and data collected from all the fine nodes (*bottom*)

Table 3.3 Comparison of the calculated mass flux and momentum flux for the fifth case from Table 3.1 at 9,000 coarse time steps, with data collected from the various grid nodes which constitute the upstream grid interface

Nodes type	Mass flux (mu/ts)	Momentum flux (mu lu/ts^2)
Coarse	−2.76074	0.04734
Overlapping fine	−2.51919	0.04569
All fine	−2.60329	0.04548

mu lattice mass unit, *lu* lattice space unit, *ts* time step

The following formula is proposed for the estimation of the time saving resulting from the migrating multiblock model:

$$Ga = \frac{N_x N_y N_z}{L_z N_x N_y + \frac{1}{m^4}(N_z - L_z) N_x N_y} \tag{3.26}$$

where N_x, N_y, and N_z are the domain length and width and height expressed in fine grids spacing respectively, and L_z is the height of the fine block. The formula is applicable only for 3D models, with the fine block covering the entire width and length. The time saving calculated by Eq. (3.26) is $Ga = 5.24$. The formula overestimated the time gain by 6 %. This could be used as an indication of the code level of efficiency.

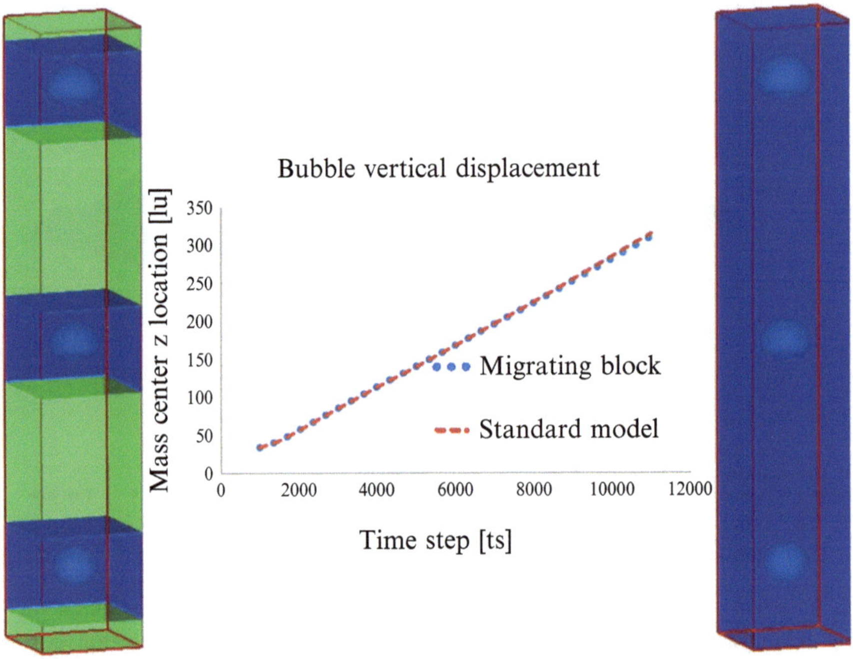

Fig. 3.25 Phase field contours and bubble vertical displacement versus time steps comparison between the migrating multiblock simulation and the standard model with $M_0 = 682$, $E_0 = 311$, and $\mathrm{Re} = 7.1$

3.3 Migrating Multiblock for the Particle-Interaction-Potential LBM

The particle-interaction-potential LBM (SC model) is a very attractive numerical tool for simulating multiphase and multicomponent flows. The SC's main advantage over other LBM models lays in its capability of handling multiphase fluids with density and viscosity contrast. The SC model has good capabilities for simulating droplets contact angle behavior, and it will be used for the study of droplet sedimentation on flat surfaces.

3.3.1 Droplet Sedimentation and Settling on a Horizontal Wall in 3D Geometry

The slow viscous motion of deformable drops is of interest in processes such as emulsion formation, mixing and separation, and droplets sprays. In the real systems these viscous motion of droplets are most of the time coupled with bounded surfaces.

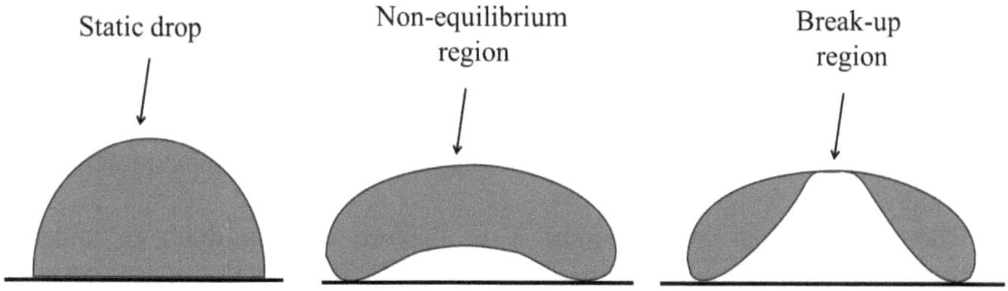

Fig. 3.26 Drops shape regimes, which include static, nonequilibrium and breakup regions

This led to study of droplet sedimentation on horizontal surfaces and the interesting physics occurring during this phenomenon. When a drop of one fluid settles through another fluid on a plane horizontal surface, it was observed that some of the external fluid is trapped in a squeeze film between the drop and the plane (Jones and Wilson 1978; Ascoli et al. 1990; Yiantsio and Davis 1990). Ascoli et al. (1990) studied the fluid drainage using a numerical scheme and observed that the droplet forms a dimple shape as it approaches the linear wall. They provided information and insight into details of fluid drainage and mechanism behind the dimple formation.

Recently, Hodges et al. (2004) studied the static shape of sessile nonwetting droplet as it sediments under the influence of gravity. They found that the increase in Bond number reduces the sphericity of the droplet. Lister et al. (2006), instead considered the sedimentation of quasi-static droplets at different Bond numbers and analyzed the drainage of fluid trapped under the droplet. A regime diagram was produced by solving a modified pressure equation for the fluid just outside the drop expressed as:

$$p = \frac{\gamma P_d}{b} - \Delta\rho g h - \frac{\gamma h_{xx}}{\left(1 + h_x^2\right)^{3/2}} \tag{3.27}$$

and solving a film evolution PDE is expressed as follows:

$$h_t + \left(\frac{h^3}{3}\left(Bh + \frac{h_{xx}}{\left(1 + h_x^2\right)^{3/2}}\right)_x\right)_x = 0 \tag{3.28}$$

where b is the undistorted drop radius; Pd is a dimensionless pressure scaled by γ/b Lister et al. (2006). h is the height, γ is the surface tension, and $\Delta\rho$ is the density excess. B in Eq. (3.28) is the Bond number, and is given as $B = \Delta\rho g b/\gamma$. In Eq. (3.27), the lengths were scaled by b and times were scaled by $\mu b/\gamma$. The subscripts x and t in Eqs. (3.27) and (3.28) represent the partial derivatives over space and time respectively.

Several regions of drop solutions were delineated and they included a sessile drop with no fluid trapped underneath the drop, quasi-equilibrium region with trapped fluid, whose extremes led to drop break up or to drainage, and a nonequilibrium region (sliding collar) (see Fig. 3.26). Lister et al. (2006) showed that the nonequilibrium region is only observed at high Bond numbers and drops remain sessile at Bond numbers below 2 approximately. It is obvious from Eq. (3.27) and (3.28) that the fluid–solid interaction was not considered in the governing equations. This case of sedimentation of droplet for various wetting characteristics of surface with the surrounding fluid has been left unexplored to our best knowledge.

The primary goal of the study is to explore the film-drainage under a sedimenting droplet, while considering the solid–fluid interaction forces between the droplet and wall surface, using a numerical approach. In addition to film drainage, this study intends to provide information and insight into the details and mechanism behind the deformation of droplet sedimenting on wetting and nonwetting walls. The problem is relevant to many applications and in particular to those dealing with hydrophobic or hydrophilic surfaces. In principle, these goals can also be approached from an experimental point or in the final stages of film dynamics via development of a complete asymptotic theory. However, the asymptotic theory can only suggest the dynamics of thin film formed under the droplet after its formation and does not provide a basis for the formation of thin film. Further, experiments have limitations to different range of solid–fluid interaction forces that can be considered. They also lack sufficient resolution to determine the velocity and pressure fields within the film of droplet sediment. Thus, a numerical approach provides the most productive basis for examining the thin film formation and drop deformation near the wetting wall and nonwetting walls. The SC multicomponent model used in this study lends itself as a convenient tool for simulating the sedimentation of viscous immiscible droplets in ambient liquids on a wall with the inclusion of the wall surface energy role due to the incorporation of the fluid–solid interaction force by Eq. (2.31). Furthermore, MMB offers computational time saving and enhanced interface resolution which is highly valued in 3D simulations.

Droplet sedimentation towards a horizontal plane is simulated in the following subsections, considering the fact that for example water condensation in stationary fuel tanks due to severe changes in ambient temperature occurs in the air layer above the free surface and the water droplets eventually land on the bottom wall due to gravity. To take into consideration the effects of the impact on the resting shape, contact angle, and film drainage underneath the droplets alongside with other factors, the droplets were seeded above the solid surface and were let to sediment from a distance. Since the focus of the study was on the settled droplet after sedimentation, the MMB was used to help saving considerable computational time and to constantly cover the droplet and its nearby surrounding liquid with finer grids.

A domain of $60 \times 60 \times 340$ measured in fine lattice units was used and it was composed of one fine block consisting of 216,000 nodes and two coarse blocks with a total of 126,000 nodes. The spacing ratio was $m = 2$, the relaxation times were $\tau_f = 0.9$ and $\tau_c = 0.7$ for the fine and coarse blocks, respectively. This yielded a kinematic viscosity $\nu = 0.133 lu^2/ts$. Kinematic viscosity ratio between the droplet and the surrounding fluid is considered to be one.

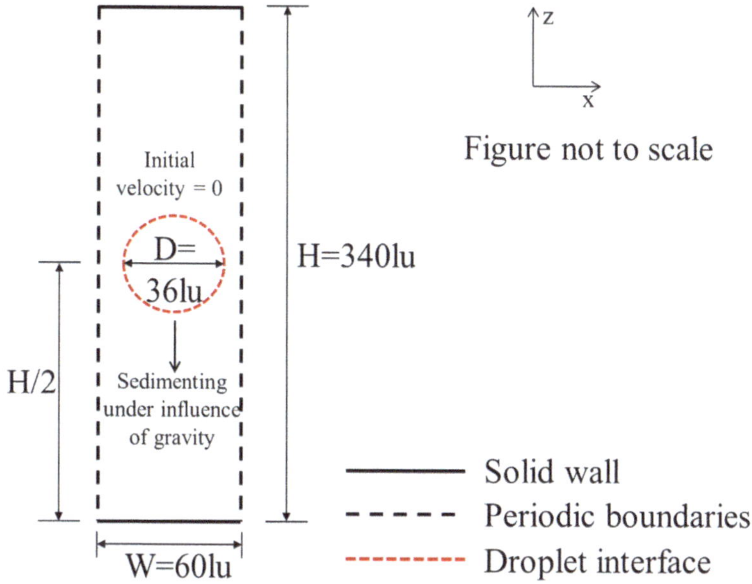

Fig. 3.27 Schematic diagram of computational domain and droplet initial condition is shown here

The suspending fluid had a density $\rho_B = 1mu/lu^3$, while the suspended fluid density was $\rho_R = 4mu/lu^3$ leading to a density ratio $\eta = 4$, which characterizes variety of immiscible liquid–liquid mixtures with small density ratio. The fluid–fluid interaction constant from Eq. (2.30) was set to $G_{\sigma\sigma'} = 0.5$. Throughout this study, kinematic viscosity ratio, density ratio, and fluid–fluid interaction constant were kept constant with the values stated above until and unless stated otherwise. Typically, multicomponent LBM models are affected by the spurious velocities. However, these spurious velocities are predominant for simulations where the density ratio between the two components is large (~100). In the present simulations, the small density ratio used for the simulation negated the spurious velocities around the interface. The authors observed the velocity vectors at the mid domain in x–z plane and found that the vector lengths of spurious velocities were either similar to flow velocities or lesser in magnitude. It was also observed that the spurious velocities do not extend farther away from the interface. Thus it was concluded that the spurious velocities do not significantly affect the physical results in the present simulations.

Schematic diagram of the initial flow condition and droplet position is shown in Fig. 3.27. A droplet of diameter D = 36 lu is placed at the center of the computational domain. It is then allowed to sediment under the influence of gravity. The initial velocity of droplet is zero. In all the simulations discussed in this paper unless specified, periodic condition was applied on the boundaries in all directions except for top and the bottom planes (i.e., in x–y plane).

Fig. 3.28 Phase field contours for four consecutive time steps, $Re_d = 9.07$ and $B = 1.66$. The migrating fine block (in *blue* (*light gray*)) moved with the droplet in (**a**) and (**b**), then the droplet moved alone in (**c**) and settled on the wall in (**d**). $Re_d \approx \frac{\overline{U}D}{\nu}$, where U is the average velocity, D is the droplet diameter, and ν is the kinematic viscosity of the droplet

3.3.2 Simulation Results and Discussions

3.3.2.1 Sessile Static Nonwetting Droplet

In order to compare the model with previous known studies, droplet static shape was investigated by letting it sediment under gravity and allowing it settle on the horizontal wall, while the wall adhesion force is theoretically zero. The fluid–solid interaction was turned off to simulate a nonwetting condition (hydrophobic surface) and to enable a comparison with the theoretical results of Hodges et al. (2004). The relationship between the fluid–solid interaction force and droplet contact angle is calculated using Dupré's law and is given by:

$$F_{adh} = \gamma(1 + \cos\theta) \tag{3.29}$$

where F_{adh} is the adhesion force per unit length, γ is the liquid surface tension, and θ is the static contact angle. Setting $F_{adh} = 0$ results in a contact angle $\theta = 180$.

The no-slip boundary condition was imposed on the lower and upper horizontal walls and a periodic condition was applied in all other directions. The migrating fine block followed the droplet mass center until the lower grid interface reached the fine grid coordinate $z = 2$ which was treated as a wall, and then the droplet was left to sediment alone until it settled on the lower wall as shown in Fig. 3.28

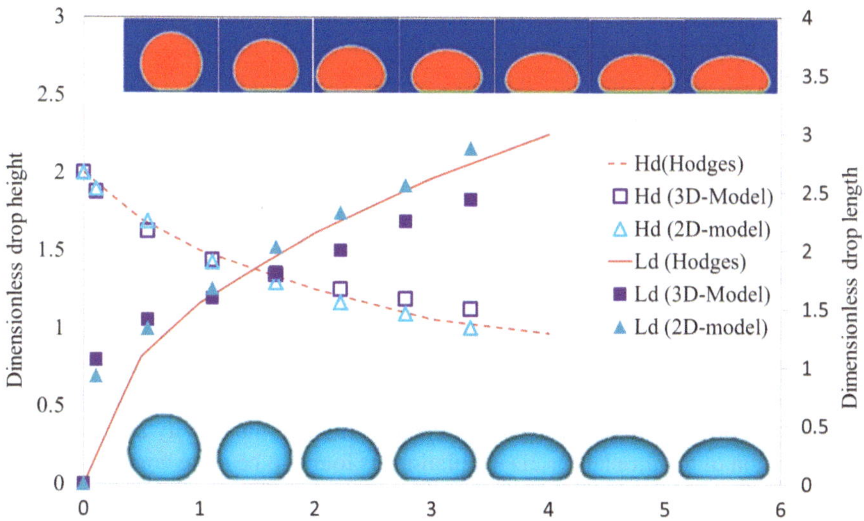

Fig. 3.29 Dimensionless droplet height and wall contact length for various Bond numbers are shown in the plot for Hodges et al. (2004) and present model. The *lower inset* show the 3D view of the droplet shapes and the upper inset is their 2D cuts. *Blue* color and *red* color in upper 2D cuts represent surrounding fluid and the droplet respectively (droplet interface is represented as a *white line* in *gray* prints of 2D upper cuts)

(see Movie 3.3.2.1—3DSCMMBsed0). Under these conditions the droplet is expected to reach a nonwetting sessile shape. In Fig. 3.28, green color indicates the coarse grid region and the blue shade indicates the region of fine grid. As the droplet moves downward, due to the influence of gravity, the fine grid region is enabled to track the center of mass of the droplet and move downward along with it. Once the droplet reaches the bottom of the surface (Fig. 3.28c), it can be observed that the fine grid region becomes stationary and further downward movement of droplet does not affect the fine grid region.

To determine the characteristics of a nonwetting drop resting on a horizontal wall, Hodges et al. (2004) solved the following dimensionless Young–Laplace relationship $P_d - BH = k$ where P_d was the droplet static pressure, B the Bond number, H the height of the drop, and k the droplet curvature. In the above mentioned Young–Laplace relationship, the length is scaled by the undistorted drop radius (b) and pressure by γ/b. Hodges et al. (2004) presented their results in a graph with the Bond number used as the independent variable.

Seven values for the gravitational constant were used in the present model to produce a variety of conditions ($0.11 \leq B \leq 3.33$) which resulted in different droplet static shapes. The uniform surface tension for Bond number was calculated by using the relation $2\gamma = \Delta PR$, where ΔP was the pressure difference measured between the fluid inside and outside of the droplet before it settled on the wall. The droplet height and wall contact length were measured and scaled by the droplet radius. The results presented in Fig. 3.29 matched reasonably well with the 2D results of Hodges et al. (2004). With the increase of Bond number, the dimensionless drop height is found to decrease and the dimensionless drop contact length is found to increase. As the Bond number increases, gravitational force dominates the surface tension force and thus resulting in the flattening of droplet. We believe that the small difference between the theoretical results of Hodges et al. (2004) and our study is due to the fact that the results of our simulations are obtained from 3D simulations, while that of Hodges et al. stem from 2D analysis. The flattening of the droplet with the increase of size seems to have been overpredicted in the theoretical analysis of Hodges et al. which is typical of the 2D results. The insets of static droplet shape at seven differently simulated Bond numbers are shown in Fig. 3.29. Flattening of the droplet with an increase in the Bond number is obviously apparent from the static droplet shapes. In order to observe the 2D droplet on the deformation of the droplet, simulations for 2D results were also performed for the same values of Bond number while maintaining the viscosity ratio of 1 and density ratio of 4. The deformation of the droplet for 2D simulations is found to be more comparable with the Hodges 2D analytical analysis. For larger Bond numbers, the droplet flattens quickly with a 2D assumption.

3.3.2.2 Sessile Static Wetting Droplet

Influence of Bond number on the shape of static droplet has previously been studied for a nonwetting surface (Hodges et al. 2004). However, most of the realistic surfaces are based on wetting properties which can influence the droplet shape. To study the influence of surface wetting on the static droplet height and wall contact length, fluid–solid interaction forces were used in the SC model and the simulations were performed for different Bond numbers.

For stationary wetting droplet resting on a horizontal wall, the contact angle is determined by Young's equation:

$$\cos\theta = \frac{\gamma_{SB} - \gamma_{SR}}{\gamma_{RB}} \tag{3.30}$$

where γ_{SB}, γ_{SR}, and γ_{RB} are the interfacial tension between the solid wall and the droplet, the solid wall and the suspending fluid, and between the two immiscible fluids, respectively. In the SC model, the interfacial tensions are directly proportional to the interaction forces of Eq. (2.30) and Eq. (2.31), and these forces are dependent on the constant which determines their magnitude.

The fluid–solid interaction constants for the surrounding fluid $G_{ads}^{B} = 0.14$, and for the droplet $G_{ads}^{R} = -0.11$ were selected to yield a wetting droplet with an acute contact angle $\theta \approx 76^0$. Three values for the gravitation constant were used in this simulation to modify the Bond number $0.55 \leq B \leq 1.66$. The drop Reynolds number is given by $\mathrm{Re}_d = {}^{\overline{U}D}/_{\nu}$, where $\overline{U}$ is the average velocity, D is the droplet diameter, and ν is the kinematic viscosity. The phase field contours of four consecutive time steps are shown in Fig. 3.30.

Fig. 3.30 Phase field contours for four consecutive time steps with $G_{ads}^{R} = -0.11$, $G_{ads}^{B} = 0.14$, $B = 1.11$, and $Re_d = 5.52$. $Re_d \approx \frac{\overline{U}D}{\nu}$, where U is the average velocity, D is the droplet diameter, and ν is the kinematic viscosity of the droplet

As in the case of nonwetting static droplet, fine grid region is found to move along with the droplet and eventually becomes stationary as the droplets reach close to the wall. A comparison of Figs. 3.28d and 3.30d suggests that the droplet shape with the wetting wall condition is more spread out and the contact angle between the droplet and wall is less than the contact angle for nonwetting droplet. With an increase of Bond number, the wetting droplet is observed to flatten out and which results in the increase of its wall contact length (Fig. 3.31). Insets in Fig. 3.31 show the static droplet shape for three different Bond numbers used in the simulations. It should be observed that the droplet contact angle was not affected by the change in the Bond number as shown in the insets of Fig. 3.31, and this was consistent with Eqs. (3.29) and (3.30) since both equations were independent of the acceleration constants which were used to vary the Bond number.

A quantitative comparison of dimensionless drop height and dimensionless wall contact length for wetting and nonwetting droplets is also presented in Fig. 3.31. The inclusion of the surface energy in wetting droplet simulations caused the droplet to spread, and by conservation of mass there was an increase in the droplet dimensionless contact length, and a decrease in the dimensionless height. The plot suggests that the flattening of droplet with an increase of Bond number follows similar trend for both wetting and nonwetting droplets, but with an increase in the wall contact length for the wetting droplets. The calculated mean values for the three Bond number cases from both the wetting with $\theta \approx 76°$ and the nonwetting simulations were compared. The results indicated an increase of 62 % for the contact length and decrease of 11 % for the drop height for the wetting cases with

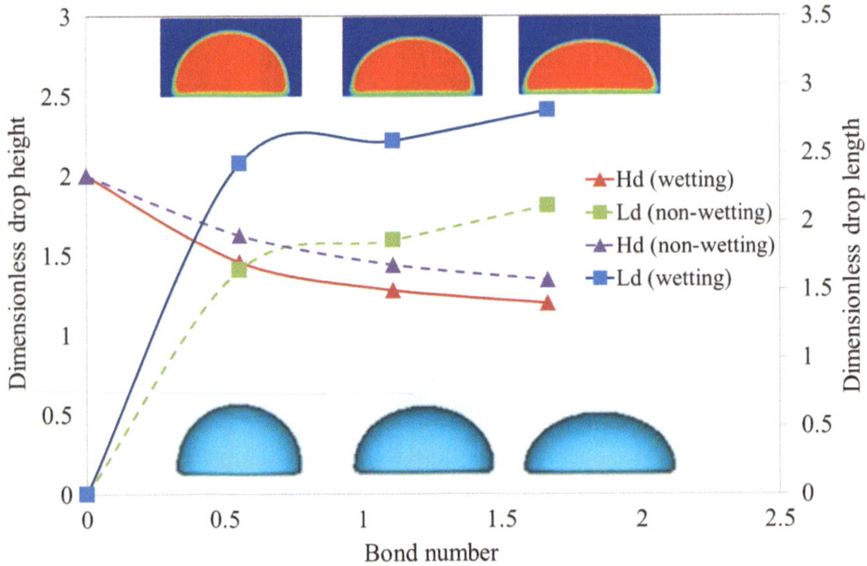

Fig. 3.31 Comparison of the dimensionless droplet height and wall contact length for various Bond numbers between the wetting and the nonwetting droplet cases. The *insets* are the phase field contours for the wetting droplets

respect to the nonwetting ones. The study of wetting and nonwetting surfaces for static droplet shape provided enough proof that the surface forces acting upon a sedimenting droplet can play a vital role in determining the spreading characteristic of the droplet on the surface.

3.3.2.3 Effects of Solid–Fluid Interaction Constants on the Drainage Underneath Wetting Droplets

As has been discussed in Chap. 1, Lister et al. (2006) studied the drainage underneath sedimenting quasi-static droplet while neglecting the solid–fluid interactions forces. However, from the previous section where a comparison of the droplet shape at different Bond numbers for static-wetting and nonwetting droplets was presented, it can be seen that the droplet shape varies considerably with the change in the surface wetting condition. Thus in this section, the fluid trapped underneath the droplet while changing the solid–fluid interaction constants in Eq. (2.31) is analyzed.

In order to study the phenomenon of solid–fluid interaction forces on sedimenting droplet, droplet sedimentation was simulated with Bond number B = 0.55 and the fluid–solid interaction constant for the droplet, $G^R_{ads} = -0.11$. The fluid–solid interaction constants for the surrounding fluid, G_{ads} ranging from 0.09 to 0.18, were used. It should be noted that the negative values of interaction parameters (Gads) produce solvophilic characteristics of surface with the liquid and

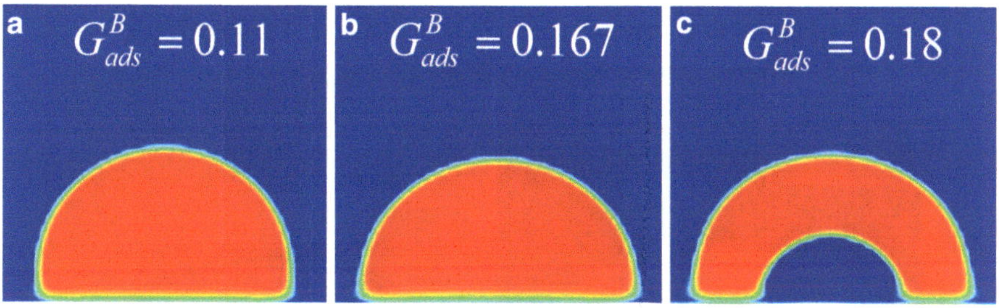

Fig. 3.32 Shape of sedimenting droplet is shown in three different regimes. B = 0.55, G_{ads}^{R} = − 0.11. (**a**) Droplet sediment shows static droplet regime with all the surrounding fluid drained from beneath the droplet at long periods of time. (**b**) Droplet forms a quasi-equilibrium shape with a thin film of surrounding fluid entrapped between the droplet and the surface, and (**c**) droplet forms a nonequilibrium shape, where large volume of surrounding fluid gets trapped under the droplet and causes the droplet to shift randomly with time (droplet interface is represented as a *white line* in *gray* prints)

the positive values of interaction parameters (Gads) produce solvophobic characteristics for surface with respect to the liquid. For G_{ads} values of 0.09–0.15, droplet behaves as sessile droplet (Fig. 3.32a) and the wetting characteristics of droplets are found to increase with increasing G_{ads} value, which is consistent with the Eq. (3.30). At $G_{ads} = 0.167$, it is observed that the surrounding fluid film gets trapped under the droplet and after long period quasi-equilibrium state (Fig. 3.32b) is reached. Buoyancy of the trapped lighter fluid and capillary forces counterbalance the droplet wall attraction leading to a stable film underneath the droplet. Most interesting phenomenon is observed when the value of $G_{ads} \geq 0.167$. The droplet forms a nonequilibrium state (Fig. 3.32c) at these conditions. Large amount of surrounding fluid gets trapped under the droplet. The force dominance in these situations favors the nonstatic film formed beneath the droplet, and this leads to droplet random shift from its initial location. As the value of G_{ads} increases, nonstatic film can lead to potential breakup of the droplet.

Figure 3.33 shows the density and the vertical velocity contours of a droplet simulated in nonequilibrium state with conditions similar to ones used for Fig. 3.32c. The instantaneous streamlines of the trapped fluids indicate the direction of the flow underneath the droplet. This flow leads to a region of higher positive velocity at the contact surface of the droplet with the wall, which destabilizes the droplet and allows a constant change in its shape and location.

In order to find the dependence of three regions on the values of G_{ads}^{R}, the simulations were performed for G_{ads}^{R} ranging from −0.09 to −0.155. The critical values of G_{ads} at which surrounding fluid is trapped beneath the droplet is shown in Table 3.1 for all values of G_{ads}^{R}. With a decrease in the values of G_{ads}^{R}, the critical value of G_{ads} at which quasi-equilibrium shape is observed also decreases (Table 3.4).

An interesting observation can be formed when the critical values of G_{ads} are closely studied from Table 3.4. The critical values are found to be colinear; the equation of line when derived from the solid–fluid interaction constants observed in

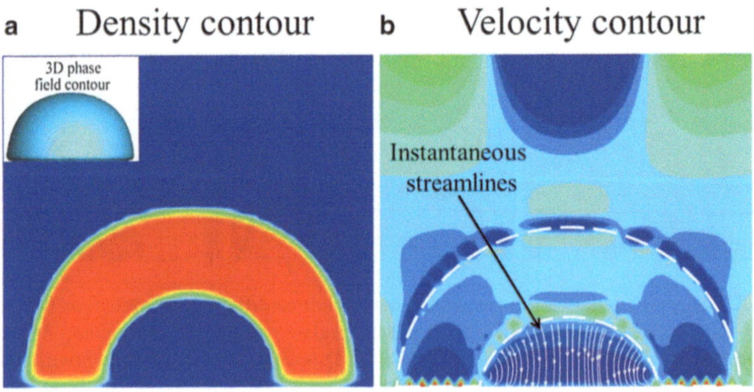

Fig. 3.33 Droplet sedimentation on a horizontal wall in the nonequilibrium region with $G_{ads}^{R} = -0.11$ and $G_{ads}^{B} = 0.18$. (**a**) Density contour, with an *inset* showing the 3D phase field contour of the drop. (**b**) Vertical velocity contour, with instantaneous local streamlines indicating the direction of the flow of the trapped ambient fluid underneath the droplet. A region of high positive velocity is observed at the droplet contact surface which leads to unstable droplet conditions (droplet interface is represented as a *white line* in *gray* prints)

Table 3.4 $G_{ads}{}^{B}$ value at which QE state is achieved in the simulations for the corresponding $G_{ads}{}^{R}$ values is tabulated here

$G_{ads}{}^{R}$	$G_{ads}{}^{B}$ − QE	QE phase
−0.090	~0.280	$\boldsymbol{G_{ads}{}^{R}} = -0.090$
−0.100	~0.220	$\boldsymbol{G_{ads}{}^{B}} = 0.280$
−0.105	~0.190	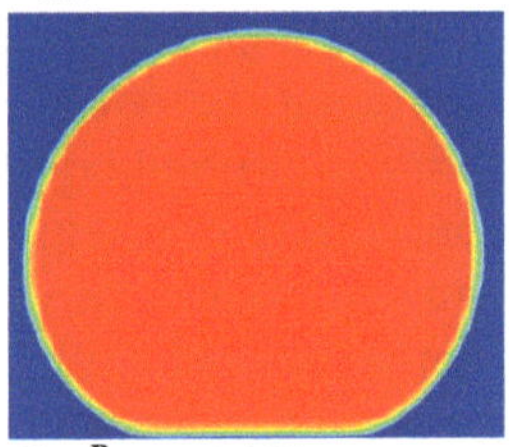
−0.110	~0.167	
−0.115	~0.138	
−0.120	~0.110	$\boldsymbol{G_{ads}{}^{R}} = -0.155$
−0.130	~0.053	$\boldsymbol{G_{ads}{}^{B}} = -0.093$
−0.140	~−0.004	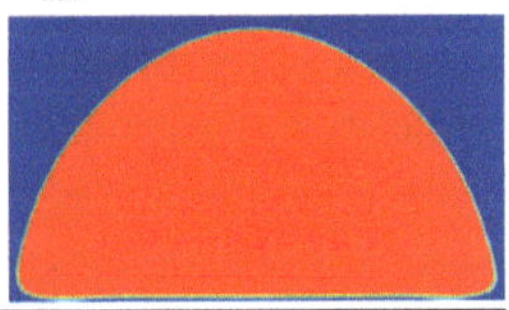
−0.150	~−0.067	
−0.155	~−0.093	

Also shown in the table is the QE state phase diagram of droplets for extreme values of $G_{ads}{}^{R}$ chosen in the present simulations (droplet interface is represented as a white line in gray prints)

the table is found to be $G_{ads}^{B} = 5.7G_{ads}^{R} + 0.794$ as shown in Fig. 3.34. The constant values corresponding to points above the line produce a nonequilibrium droplet shape during sedimentation, while the points below the line produce a static shape of droplet during the sedimentation. In SD region, decrease in the values of G_{ads}^{R}

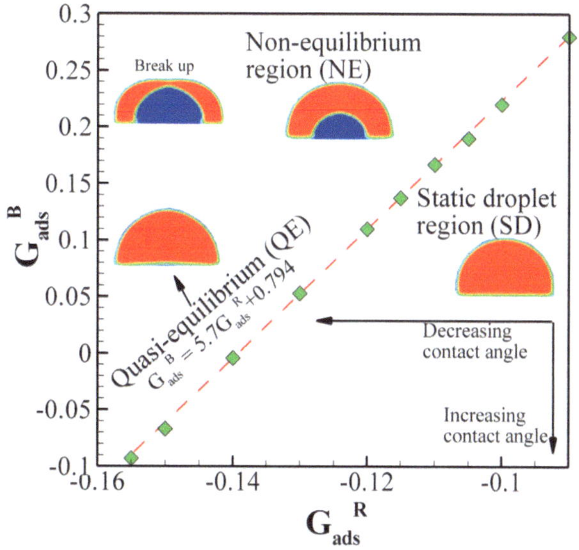

Fig. 3.34 Phase diagram of shape of sedimenting droplet for different values of G_{ads} and G^{R}_{ads}. The *dashed line* (*red line*) is the theoretical estimate of QE line for the present data, and the *solid symbols* represent the simulated cases in this study (droplet interface is represented as a *white line* in *gray* prints)

corresponds with a decrease in contact angle or in other terms increases the wetting phenomenon; decrease in the values of G_{ads} increases the contact angle of the droplet. In the region of NE, increasing values of G_{ads} for a constant value of G^{R}_{ads} cause the droplet to breakup (Fig. 3.34).

3.3.2.4 Computational Time Saving

For the estimation of the time gain ratio resulting from the use of MMB in the 3D simulations, the following formula was used:

$$Ga = C_c \frac{N_x N_y N_z}{L_z N_x N_y + \frac{1}{m^4}(N_z - L_z) N_x N_y} \tag{3.31}$$

where N_x N_y and N_z are the domain length and width and height respectively expressed in fine grids spacing, and L_z is the height of the fine block. The formula is applicable only to 3D models, with the fine block covering the entire width and length. A time gain ratio $Ga \approx 4.17$ was calculated for the above presented simulations using Eq. (3.31) for $L_z = 60$ and estimated $C_c = 0.95$.

Chapter 4
Hybrid LBM for Surfactant-Covered Droplets

Abstract This chapter describes a novel hybrid lattice Boltzmann model for the study of the droplet flow behavior in an immiscible medium with insoluble nonionic surfactant adhering to its interface. The evolution of the surfactant concentration on the interface is modeled by the time-dependent surfactant convection-diffusion equation and solved by a finite difference scheme. The fluid velocity field, the pressure, and the interface curvature are calculated using the lattice Boltzmann method (LBM) for binary fluid mixtures. The coupling between the LBM and the finite difference scheme is achieved through the LBM macroscopic variables and the surfactant equation of state. The Gunstensen LBM is used here because it provides local and independent application of a distinct interfacial tension on the individual nodes of the droplet interface. The hybrid model is developed and successfully applied to droplets deformations and surfactant distribution under simple shear flow, uniaxial flow, and buoyancy.

This chapter aims at adding a humble contribution to the wealth of several numerical methods proposed by some devoted researchers for the study of surfactant covered drop's behavior under diverse flow conditions. The work represents an expansion of the Gunstensen model applicability to cover the study of immiscible droplets with nonionic insoluble surfactant adhering to their interfaces. Adding the surfactant effects on the Gunstensen model is facilitated by the ease in the initialization and the tracking of the surfactant concentration on the interface, which location is well defined in the LBM model at each time step, and by the independent local application of a nonisotropic interfacial tension on the droplet external surface.

Electronic supplementary material Supplementary material is available in the online version of this chapter at http://dx.doi.org/10.1007/978-1-4899-7402-0_4. Videos can also be accessed at http://www.springerimages.com/videos/978-1-4899-7401-3

H. Farhat et al., *Accelerated Lattice Boltzmann Model for Colloidal Suspensions: Rheology and Interface Morphology*, DOI 10.1007/978-1-4899-7402-0_4,

4.1 Surfactants Convection-Diffusion Equation

The surfactant concentration distribution on the interface of an immiscible mixture is governed by the following dimensional form of the general time-dependent surfactant convection-diffusion equation (Milliken and Leal 1991):

$$\partial_t\Gamma + \nabla_s \cdot (\boldsymbol{u}_s\Gamma) + k\Gamma u_n = D_s\nabla_s^2\Gamma + q_{chem} + q_f \tag{4.1}$$

In Eq. (4.1) $\partial_t\Gamma$ accounts for the temporal change in the interface surfactant concentration, $\nabla_s \cdot (\boldsymbol{u}_s\Gamma)$ is the convection term, and $k\Gamma u_n$ models the effects of the change in the surface curvature on the surfactant concentration distribution. $D_s\nabla_s^2\Gamma$ is the diffusion term, q_{chem} accounts for the interface surfactant formation due to chemical reaction, and q_f accounts for the net flux to the interface from the bulk phases due to adsorption–desorption (both q_{chem} and q_f effects are not considered in this work).

The objective here is to rewrite Eq. (4.1) as a function of the LBM variables in 2D geometry. The units used in this work are identified as follows: spatial lattice unit [*lu*], time step [*ts*], mass unit [*mu*], lattice mole [*lmol*].

The following is pseudocode for initialization and averaging out the surfactant concentration at the interface

```
DOFOR x = 0, XLENGTH
 DOFOR y = 0, YLENGTH
   IF interface_x,y = 1 and timestep = 0
     Γ_x,y = Initial value
   ELSEIF interface_x,y = 1 and timestep > 0
     Count = 0
     Ccount = 0
    DOFOR p = x-3, x+4
      DOFOR q = y-3,y+4
        IF Γ_p,q != 0
          Count = Count +1
          Ccount = Count + Γ_p,q
        ENDIF
       END DO
      END DO
    Γ_x,y = Ccount/Count
   ENDIF
  END DO
END DO
```

For insoluble surfactant, subjected to flow conditions, in which the convection time scale is much greater than the diffusion time scale, the time-dependent convection-diffusion equation is reduced to the following form:

$$\partial_t \Gamma + \nabla_s \cdot (\boldsymbol{u}_s \Gamma) + k\Gamma u_n = 0 \tag{4.2}$$

where ∇_s is the surface gradient, Γ is the surfactant concentration, k is the curvature calculated by Eq. (2.17), and u_n is the normal velocity magnitude at the interface and it is given by:

$$u_n = \boldsymbol{u} \cdot \boldsymbol{n} = u_x n_x + u_y n_y \tag{4.3}$$

where $\boldsymbol{u}$ is the macroscopic velocity derived from Eq. (2.9). $\boldsymbol{u}_s$ is the tangential velocity with vertical and horizontal component magnitudes expressed respectively as:

$$\begin{aligned} u_{sx} &= u_x - n_x^2 u_x - n_x n_y u_y \\ u_{sy} &= u_y - n_y^2 u_y - n_x n_y u_x \end{aligned} \tag{4.4}$$

The following is pseudocode for implementing the tangential velocity in Gunstensen-based LBM

```
DOFOR x = 0, XLENGTH
  DOFOR y = 0, YLENGTH
    IF interface_x,y = 1
       Usx_x,y = ux_x,y – f_x^2*ux_x,y – f_x*f_y*uy_x,y
       Usy_x,y = uy_x,y – f_y^2*uy_x,y – f_x*f_y*ux_x,y
  END DO
END DO
```

Using the product rule $\nabla_s \cdot (\boldsymbol{u}_s\Gamma)$ can be expressed as a function of the tangential velocity and the normal to the interface components, respectively. The term $k\Gamma u_n$ is straightforwardly derived as the multiplication of three scalar quantities. Combining all the terms of Eq. (4.2) leads to the following simplified equation:

$$\partial_t\Gamma + C_1\partial_x\Gamma + C_2\partial_y\Gamma + C_3\Gamma = 0 \tag{4.5}$$

where the coefficients C_j are expressed as follows:

$$\begin{aligned} C_1 &= u_{sx} \\ C_2 &= u_{sy} \\ C_3 &= \left(kn_x u_x + kn_y u_y\right) + n_y^2\partial_x u_{sx} + n_x^2\partial_y u_{sy} - n_x n_y\left(\partial_y u_{sx} + \partial_x u_{sy}\right) \end{aligned} \tag{4.6}$$

When the diffusion of the interface surfactant is considered, the surfactant concentration time-dependent convection-diffusion equation is given by the following:

$$\partial_t\Gamma + \nabla_s \cdot (\boldsymbol{u}_s\Gamma) + k\Gamma u_n = D_s\nabla_s^2\Gamma \tag{4.7}$$

D_s is the surface diffusion constant which can be determined in lattice units $[lu^2ts^{-1}]$ from the following relationship: $Pe_s = \dot{\gamma}R_0^2/D_s$, where Pe_s is the surface Pèclet number which represents the ratio between the convection and diffusion of the surfactants on the interface and R_0 is the droplet radius. The final form of the equation is given by:

$$\partial_t\Gamma + C_1\partial_x\Gamma + C_2\partial_y\Gamma + C_3\Gamma + C_4\partial_{xx}^2\Gamma + C_5\partial_{yy}^2\Gamma + C_6\partial_{xy}^2\Gamma = 0 \tag{4.8}$$

where the coefficients C_j are calculated by the following:

$$\begin{aligned} C_4 &= \left(n_x^2 - 1\right)D_s \\ C_5 &= \left(n_y^2 - 1\right)D_s \\ C_6 &= 2n_x n_y D_s \end{aligned} \tag{4.9}$$

The following is pseudocode for a calculation of tangential derivatives and constants of Eq. (4.8)

```
DOFOR x = 0, XLENGTH
  DOFOR y = 0, YLENGTH
    IF interface_x,y = 1
      DOFOR i = 1, Q
        n_x = next_x_x[i], n_y = next_y_y[i]
        nn_x = next_x_nx[i], nn_y = next_y_ny[i]
        nnn_x = next_x_nnx[i], nnn_y = next_y_nny[i]
        XDer_usx_x,y = XDer_utangx_x,y + weighting_factor * ( 4.5 * usx_n_x,n_y - 0.9 *
        usx_nn_x,nn_y + 0.1 * usx_nnn_x,nnn_y ) * cx_[i];
        XDer_usy_x,y = XDer_usy_x,y + weighting_factor * ( 4.5 * usy_n_x,n_y - 0.9 * usy_nn_x,nn_y
        + 0.1 * usy_nnn_x,nnn_y ) * cx_[i];
        YDer_usx_x,y = YDer_usx_x,y + weighting_factor * ( 4.5 * usx_n_x,n_y - 0.9 *
        usx_nn_x,nn_y + 0.1 * usx_nnn_x,nnn_y ) * cy_[i];
        YDer_usy_x,y = YDer_usy_x,y + weighting_factor * ( 4.5 * usy_n_x,n_y - 0.9 * usy_nn_x,nn_y
        + 0.1 * usy_nnn_x,nnn_y ) * cy_[i];
      END DO
      C1_x,y = usx_x,y
      C2_x,y = usy_x,y
      C3_x,y = -( Kurv_x,y * ux_x,y* f_x_x,y + Kurv_x,y * uy_x,y* f_y_x,y) - f_x_x,y*f_y_x,y*(YDer_usx_x,y +
          XDer_usy_x,y)+ f_y_x,y * f_y_x,y*XDer_usx_x,y+ f_x_x,y * f_x_x,y*YDer_usy_x,y;
      C4_x,y = (f_x_x,y*f_x_x,y-1)*Ds;
      C5_x,y = (f_y_x,y* f_y_x,y -1)*Ds;
      C6_x,y = (f_x_x,y * f_y_x,y)*2*Ds;

    END IF
  END DO
END DO
```

The surfactant concentration effect on the interfacial tension of the droplet can be imposed by either one of the surfactant equations of state; the Adamson linear equation (Velankar et al. 2004):

$$\sigma_s = \sigma_0 - \Gamma RT \tag{4.10}$$

or the Langmuir nonlinear equation (Eggleton et al. 2001):

$$\sigma_s = \sigma_0 + RT\Gamma_\infty \ln\left(1 - \frac{\Gamma}{\Gamma_\infty}\right) \tag{4.11}$$

where σ_s is contaminated interfacial tension, R is the universal gas constant, and T is the temperature in Kelvin. The two equations can be rewritten for convenience as follows:

$$\sigma_s = \sigma_0(1 - E_0\Gamma^*) \tag{4.12}$$

$$\sigma_s = \sigma_0[1 + E_0 \ln(1 - \Gamma^*)] \tag{4.13}$$

where σ_0 is the surface tension of a clean droplet, and E_0 the surfactant elasticity. E_0 is a positive dimensionless parameter which determines the strength of the surfactant concentration effects on the droplet interfacial tension and it is given by (Velankar et al. 2004):

$$E_0 = \frac{\Gamma_\infty RT}{\sigma_0} < 1 \tag{4.14}$$

The dimensionless surfactant concentration is calculated by the following ratio:

$$\Gamma^* = \frac{\Gamma}{\Gamma_\infty} \tag{4.15}$$

where Γ_∞ is the saturation surfactant concentration which can be derived from Eq. (4.14) in lattice units as $\Gamma_\infty = \frac{E_0\sigma_0}{RT}\,[lmol/lu^2]$ and the product $RT = 1/3$ is used for the isothermal LBM.

The following is pseudocode for altering the interfacial tension based on local surfactant concentration using the Langmuir equation of state

```
DOFOR x = 0, XLENGTH
   DOFOR y = 0, YLENGTH
      IF interface_x,y = 1
         σ_x,y = σ_initial*(1+ E0*log( 1 – Γ_x,y/Γ_∞);
      ELSE
         σ_x,y = 0
      END IF
   END DO
END DO
```

4.2 The Hybrid LBM Model

The presented model uses the Gunstensen LBM for the calculation of the flow pressure, the velocity field, and for tracking the fluid–fluid interface. During initialization of the LBM, and after locating the interface's nodes through the magnitude of the phase field gradient $\nabla\rho^N$ of Eq. (2.16), an initial surfactant concentration Γ_i is imposed on the interface with a controllable thickness as shown in Fig. 4.1c. The selected value of Γ_i should produce the desired surfactant coverage $c_{in} = \Gamma_i/\Gamma_\infty$ needed for the particular case.

The calculated velocity component magnitudes (u_x, u_y), the droplet curvature, and the interface normal component magnitudes (k, n_x, n_y) are then used for the derivation of the simplified surfactant-diffusion equation, which is eventually solved by a finite difference scheme resolved on the same spatial lattice grid. The hopscotch explicit and unconditionally stable finite difference scheme (Tanehill et al. 1997) is used here. This scheme uses two consecutive sweeps through the domain. For the first sweep $\Gamma_{i,j}^{n+1}$ is calculated at each grid point, for which $i + j + n$ is even, by a simple explicit scheme:

$$\begin{aligned}&\frac{\Gamma_{i,j}^{n+1}-\Gamma_{i,j}^{n}}{\Delta t}+C_1\frac{\Gamma_{i+1,j}^{n}-\Gamma_{i-1,j}^{n}}{2\Delta x}+C_2\frac{\Gamma_{i,j+1}^{n}-\Gamma_{i,j-1}^{n}}{2\Delta y}+C_3\Gamma^n\\&+C_4\frac{\Gamma_{i+1,j}^{n}-2\Gamma_{i,j}^{n}+\Gamma_{i-1,j}^{n}}{\Delta x^2}+C_5\frac{\Gamma_{i,j+1}^{n}-2\Gamma_{i,j}^{n}+\Gamma_{i,j-1}^{n}}{\Delta y^2}\\&+\frac{C_6}{2\Delta x}\left(\frac{\Gamma_{i+1,j+1}^{n}-\Gamma_{i+1,j-1}^{n}}{2\Delta y}-\frac{\Gamma_{i-1,j+1}^{n}-\Gamma_{i-1,j-1}^{n}}{2\Delta y}\right)=0\end{aligned}\tag{4.16}$$

For the second sweep $\Gamma_{i,j}^{n+1}$ is calculated at each grid point, for which $i + j + n$ is odd, by a simple apparent implicit scheme:

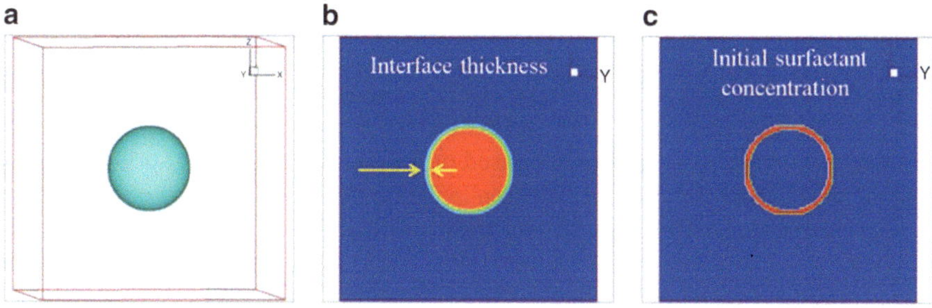

Fig. 4.1 Initialization of the surfactant concentration at the interface of a 3D droplet: (**a**) 3D domain with a central droplet, (**b**) 2D view of the phase field showing the interface thickness, (**c**) 2D view of the uniform surfactant concentration contour on the interface

$$\frac{\Gamma_{i,j}^{n+1}-\Gamma_{i,j}^{n}}{\Delta t}+C_1\frac{\Gamma_{i+1,j}^{n+1}-\Gamma_{i-1,j}^{n+1}}{2\Delta x}+C_2\frac{\Gamma_{i,j+1,k}^{n+1}-\Gamma_{i,j-1,k}^{n+1}}{2\Delta y}$$
$$+C_4\frac{\Gamma_{i+1,j}^{n+1}-2\Gamma_{i,j}^{n+1}+\Gamma_{i-1,j}^{n+1}}{\Delta x^2}+C_5\frac{\Gamma_{i,j+1}^{n+1}-2\Gamma_{i,j}^{n+1}+\Gamma_{i,j-1}^{n+1}}{\Delta y^2} \tag{4.17}$$
$$+C_3\Gamma^{n+1}+\frac{C_6}{2\Delta x}\left(\frac{\Gamma_{i+1,j+1}^{n+1}-\Gamma_{i+1,j-1}^{n+1}}{2\Delta y}-\frac{\Gamma_{i-1,j+1}^{n+1}-\Gamma_{i-1,j-1}^{n+1}}{2\Delta y}\right)=0$$

The second sweep is expressed in an implicit form, but it is solved as an explicit equation, because the first sweep provides the necessary information needed for the computation of $\Gamma_{i,j}^{n+1}$. The truncation error for the used hopscotch scheme in the presented model is of the following order $O[\Delta t, (\Delta x)^2, (\Delta y)^2]$.

Besides the fact that the hopscotch scheme is unconditionally stable, it is also simple for coding especially if the time step for the finite difference is to be modified from that of the LBM. This can be achieved by using the three-point Lagrangian interpolation for the calculation of the required coefficients C_j at a fraction of the LBM time step as follows:

$$\underbrace{C_j^{t_3-r/n}}_{1\le r<n}=\sum_{p=1}^{3}\left[C_j^{t_p}\prod_{q=1,p\neq q}^{3}\frac{(t_3-r/n)-t_q}{t_p-t_q}\right] \tag{4.18}$$

A ratio $n = t_{LBM}/t_{FD} = 4$ was used in this model unless otherwise mentioned, and r is a positive integer $1 \le r < n$. The coupling of the finite difference scheme with the LBM is realized through the surfactant equation of state. In this model the nonlinear equation of state was more often used in the simulations.

Halliday et al. (2007) derived the following relationship between the pressure jump across the interface and the surface tension parameter:

$$\Delta P=\left|\int_{P_B}^{P_R}\boldsymbol{F}d\boldsymbol{n}\right|=\alpha k \tag{4.19}$$

where $\boldsymbol{F}$ is the force from Eq. (2.16) and $\boldsymbol{n}$ is the interface normal, k is the curvature from Eq. (2.17), and P_B, P_R are the measured pressures outside and inside the droplet, respectively. This suggests based on Laplace's law for the surface tension that the magnitude of the surface tension in the model is equal to that of the surface tension parameter α from Eq. (2.16); therefore Eq. (4.13) can be reevaluated as follows:

$$\alpha=\alpha_0[1+E_0\ln(1-\Gamma^*)] \tag{4.20}$$

where α_0 is the surface tension parameter for a clean droplet. The surface tension parameter in the presented model is thus nonisotropic, and it rather changes locally based on the outcome of Eq. (5.20), which is mainly dependent on the calculated local surfactant concentration by Eq. (5.8).

No upper bond on the surfactant concentration is required in this model. An important factor which prevents any further buildup of the concentration is the Marangoni stress which is expressed as follows (Hu and Lips 2003):

$$-\nabla_s \sigma_s = -\partial_\Gamma \sigma_s \cdot \nabla_s \Gamma \tag{4.21}$$

The partial derivative $\partial_\Gamma \sigma$ where σ is expressed by Eq. (4.13) yields the following equation:

$$\partial_\Gamma \sigma_s = -\frac{RT}{(1 - \Gamma/\Gamma_\infty)} \tag{4.22}$$

Equations (4.21) and (4.22) indicate that an increase in the surfactant surface concentration leads to an increase in the Marangoni stress, which in turn slows down the surface velocity and hampers any further buildup of surfactant towards the regions of higher concentration.

The flow chart for the hybrid LBM for surfactant covered droplets is presented in Fig. 4.2.

4.3 Simulation Results and Discussions

To demonstrate the presented model suitability as a tool for investigating the surfactant-covered droplet behavior under diverse flow conditions, the model was used for the study of the surfactant effects on the droplet deformation in simple shear flow, uniaxial extensional flow, and on the terminal velocity of a buoyant surfactant-covered droplet.

4.3.1 Algorithm Validation and Testing

To validate the proposed algorithm, the results of the hybrid LBM for extensional flow in 2D domain were compared with the numerical findings of Feigl et al. (2007), who presented their results for the deformation index $D = (L - B)/(L + B)$, dimensionless major axis half length L/a, where a was the droplet radius and B was the minor axis half length, and the dimensionless maximum and minimum surfactant concentrations with respect to the equilibrium capillary number $Ca^* = Ca/(1 - E_0)$. The surfactant elasticity was determined as $E_0 = \Gamma_0 RT/\sigma_0$ in which Γ_0 was the initial concentration. The linear equation of state was used with $\Gamma^* = \Gamma/\Gamma_0$. The simulation was executed for one droplet under extensional flow with viscosity ratio $\lambda = 1$, surfactant elasticity $E_0 = 0.1$, and the

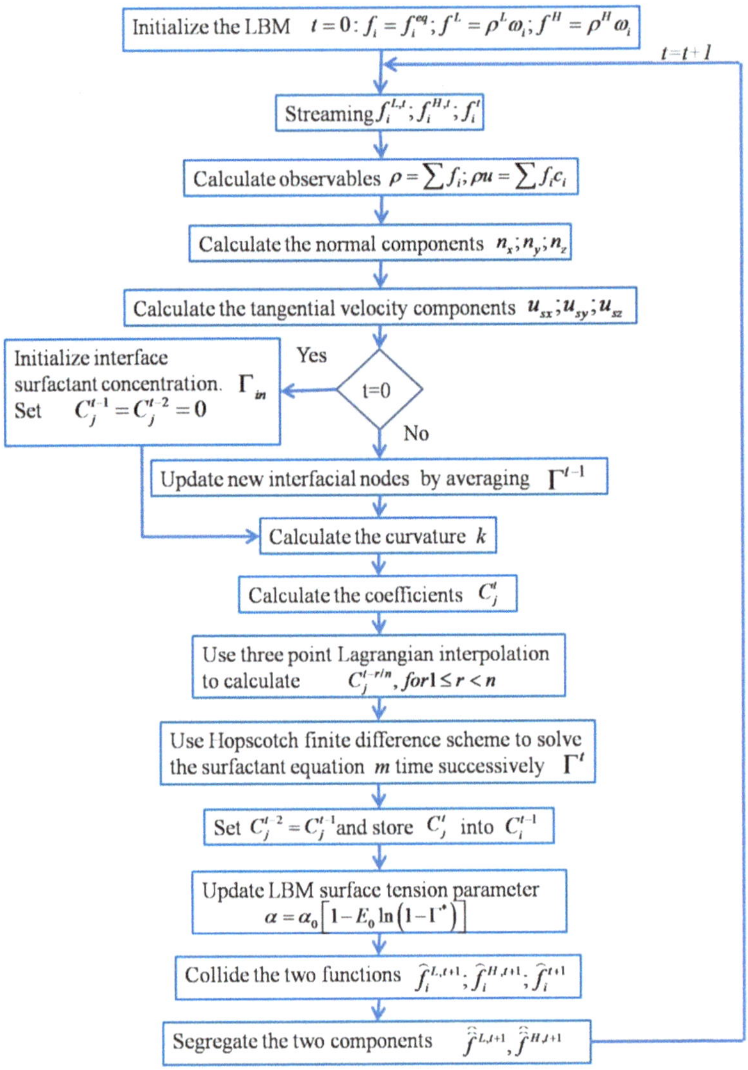

Fig. 4.2 Flow chart for the hybrid LBM for surfactant-covered droplets

coefficient $\alpha = Pe_s/Ca = 10.0$. To match the same conditions in the hybrid LBM the surface tension parameter was set to $\alpha_0 = 10^{-3}$, the initial surfactant concentration to $\Gamma_i = 3 \times 10^{-4}[lmol/lu^2]$ and the diffusion constant to $D_s = 6 \times 10^{-3}$ $[lu^2 ts^{-1}]$, for a range of capillary numbers $0.0243 \le Ca \le 0.099$. The domain was made of $123 \times 123[lu^2]$, the density was $\rho = 2.0[mu/lu^2]$, and the relaxation time was $\tau = 1.0$. The linear equation of state was used in this simulation. The results of the presented model for the droplet deformation are presented in Fig. 4.3. A comparison of the hybrid LBM results for the normalized maximum and minimum

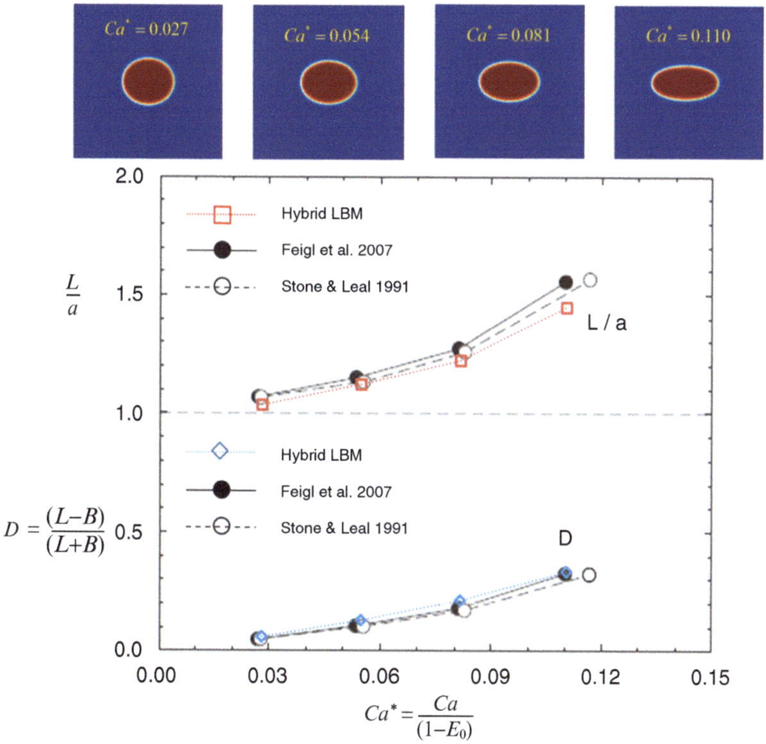

Fig. 4.3 Comparison of the hybrid LBM results for the normalized major axis half length (*upper*) and for the deformation index with the results of Feigl et al. (2007) for a drop in extensional flow. The viscosity ratio is $\lambda = 1.0$, the surfactant elasticity is $E_0 = 0.1$, and the diffusion constant is $D_s = 6 \times 10^{-3}$ corresponding to the value of $\alpha = 10.0$ from Feigl et al. (2007). The *insets* in the figure are for the droplet phase field contours (reprinted with permission)

surfactant concentration with the results of Feigl et al. (2007) is presented in Fig. 4.4.

The resilience of the presented model was tested using surfactant-covered droplet breakup under extensional flow in 2D $181 \times 81[lu^2]$ domain. The nonlinear equation of state Eq. (4.13) was applied, for viscosity ratio $\lambda = 0.143$ which was achieved by using fluid relaxation time $\tau_f = 1.21$ and a droplet relaxation time $\tau_d = 0.6$. The shearing rate was set to $\dot{\gamma} = 1.975 \times 10^{-5}[ts^{-1}]$, the surfactant elasticity to $E_0 = 0.5$, the surfactant coverage to $c_{in} = 0.2$, the Péclet number to $Pe_s = 75$, and the surface tension was $\alpha_0 = 10^{-4}$. The phase field, the surfactant concentration, and the surface tension parameter contours are shown in Fig. 4.5. It is clear from the contours that the regions of higher surfactant concentration are associated with lower surface tension, and that after breakup the two resulting droplets start to contract which reduces the dilution of the surfactant concentration.

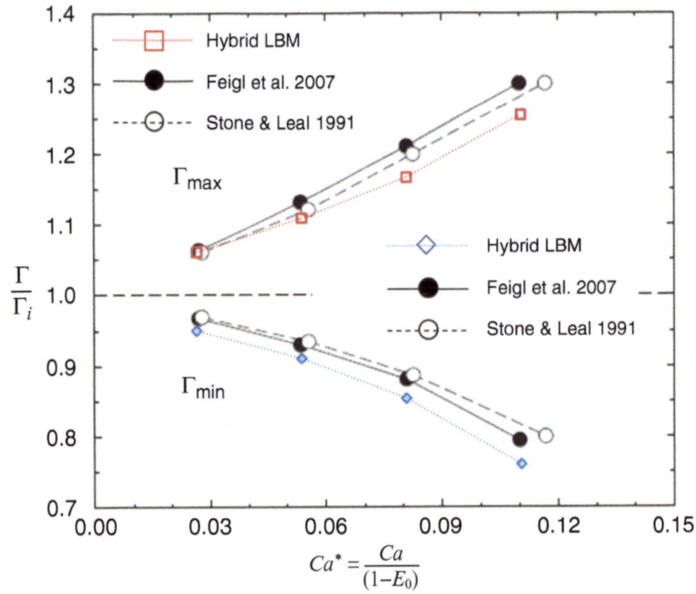

Fig. 4.4 Comparison of the hybrid LBM results for the normalized maximum and minimum surfactant concentration with the results of Feigl et al. (2007) for a drop in extensional flow. The viscosity ratio is $\lambda = 1.0$, the surfactant elasticity is $E_0 = 0.1$, and the diffusion constant is $D_s = 6 \times 10^{-3}$ corresponding to the value of $\alpha = 10.0$ from Feigl et al. (2007) (reprinted with permission)

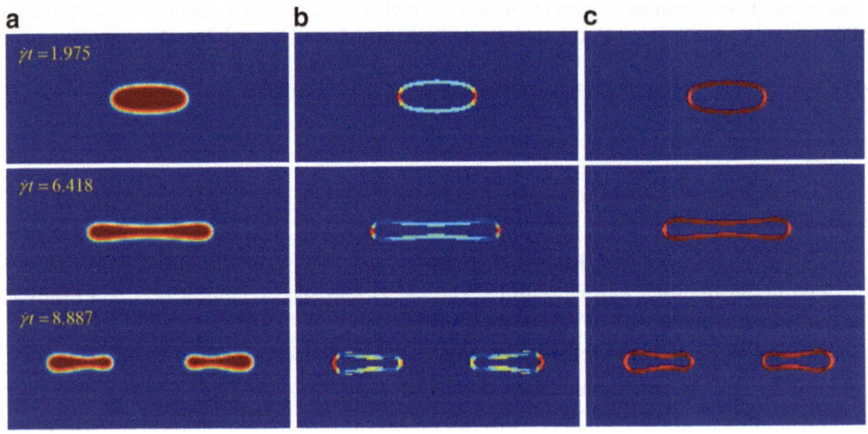

Fig. 4.5 Surfactant-covered droplet breakup in 2D domain. (**a**) Phase field contours, (**b**) surfactant concentration contours, and (**c**) surface tension parameter contours for three dimensionless time steps. The viscosity ratio is $\lambda = 0.147$, the surfactant elasticity is $E_0 = 0.5$, the surface coverage is $c_{in} = 0.2$, and the Péclet number is $Pe_s = 75$

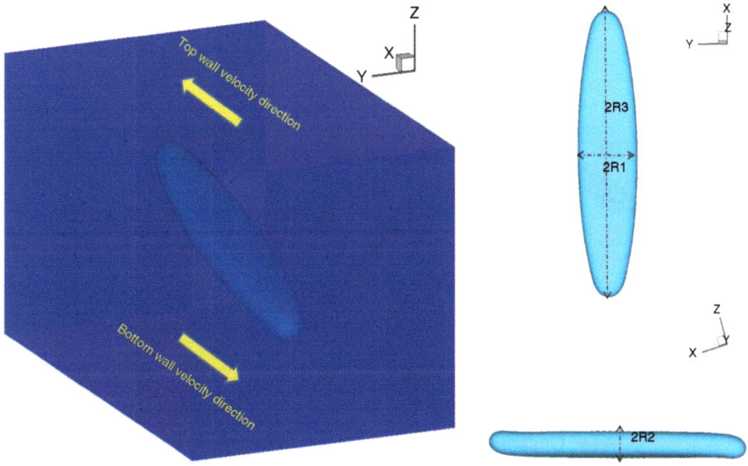

Fig. 4.6 Illustration of the LBM simulation domain with a central clean droplet under simple shear stress and the three characteristic radii used in the analysis of the results

4.3.2 Surfactant-Covered Droplets in Simple Shear Flows

To study the effects of surfactants on the droplet behavior in simple shear flows, the setting of the base model parameters was decided through studying the flow deformation characteristics of a surfactant-free droplet ($\Gamma_i = 0$). The numerical results were then compared with the experimental work of Cristini et al. (2002), who investigated the transient deformation of clean droplets in dilute emulsions for large values of the capillary numbers. Cristini et al. (2002) studied the transient lamellar microstructures of some polymer blends which exhibited elongation and flattening of the droplets under strong shear flow conditions and for low viscosity ratio. Such blend morphologies were attractive because they caused a reduction in the permeability of certain blends to hydrocarbon and to oxygen, and increased their impact strength (Cristini et al. 2002).

The droplet lamellar structure shown in Fig. 4.6 is characterized by the normalized length R_1/R_0, thickness R_2/R_0, and width R_3/R_0, where R_0 is the initial droplet radius. The interfacial area generation during the droplet deformation is calculated by R_1R_3/R_0^2.

A domain consisting of $123[lu^3]$ was used in this simulation with a suspended central clean droplet having a radius $R_0 = 20[lu]$. The interfacial tension parameter was $\alpha_0 = 2 \times 10^{-4}$, and a shear rate $\dot{\gamma} = 9.75 \times 10^{-5}[ts^{-1}]$ was imposed through moving the upper and lower walls in the directions shown in Fig. 4.6 by the following:

$$u^{\infty}(x) = \begin{pmatrix} 0 & 0 & \dot{\gamma} \\ 0 & 0 & 0 \\ 0 & 0 & 0 \end{pmatrix} \cdot x \tag{4.23}$$

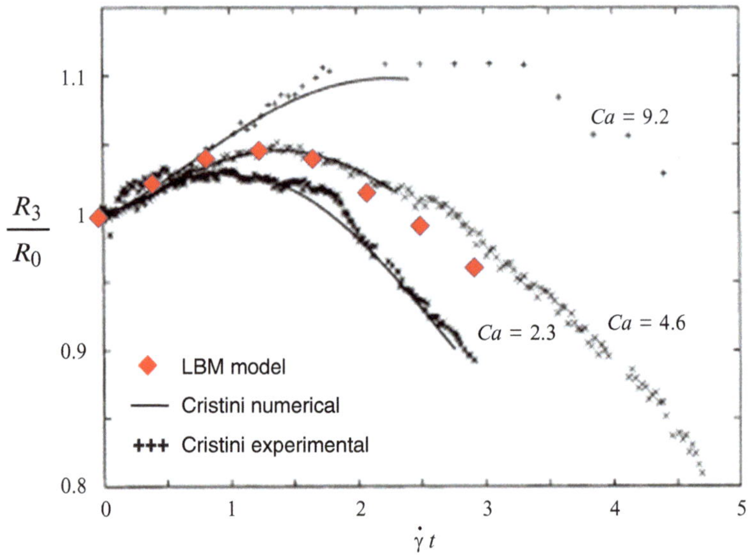

Fig. 4.7 Comparison of the presented numerical model results with the experimental and numerical results of Cristini et al. (2002) for a clean droplet dimensionless width as a function of the dimensionless time. The viscosity ratio is $\lambda = 0.1$, and the capillary number is $Ca = 4.6$ (reprinted with permission)

The periodic boundary condition was used in all other directions. The relaxation time for the ambient fluid was $\tau_m = 1.213$ and for the droplet $\tau_d = 0.571$ leading to a viscosity ratio $\eta = 0.1$. The interface viscosity was calculated by Eq. (2.15). The density of both fluids was set to $\rho = 2[mu/lu^3]$.

The droplet deformation in simple shear flows is characterized by the capillary number which is the ratio of the droplet deforming shear stress and the restoring stress due to the interfacial tension:

$$Ca = \frac{R_0 \mu_m \dot{\gamma}^m}{\sigma_0} \tag{4.24}$$

where σ_0 is the interfacial tension of a clean droplet and μ_m is the dynamic viscosity of the matrix. Equation (4.24) yielded a capillary number $Ca = 4.6$ in correspondence with one of the experimental condition of Cristini et al. (2002). The resulting dimensionless width of the droplet from the presented model with respect to the dimensionless time is shown in Fig. 4.7.

The clean droplet case showed a good agreement with the experimental data for the dimensionless time $\dot{\gamma}t \leq 2.0$.Therefore the investigation of the area generation due to the presence of surfactants will be limited to values of the dimensionless time $\dot{\gamma}t < 2.0$ while other droplet flow deformation characteristics will be discussed for time step $\dot{\gamma}t = 3.12$ corresponding to the end of the simulation time which was dictated by the desire of not allowing the droplet to deform beyond the periodic boundaries.

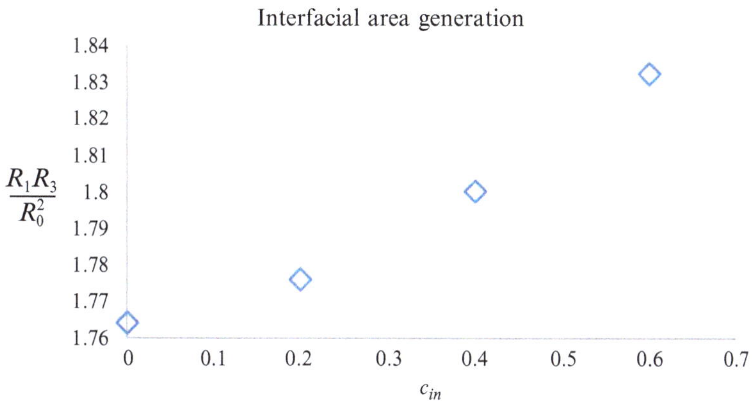

Fig. 4.8 Interfacial area generation for droplets in simple shear flow with respect to initial surfactant coverage presented at a dimensionless time step $\dot{\gamma}t = 1.17$ and capillary number $Ca = 4.6$

Table 4.1 Transient dimensionless length, percentage elongation, and reference angle of inclination measured in degrees, for the clean droplet and for the droplets with three initial values of surfactant coverage c_{in} at dimensionless time step $\dot{\gamma}t = 3.12$

C_{in}	R_1/R_0	%	$\theta°$
0	2.66	100.0	20.30
0.2	2.71	101.9	19.67
0.4	2.78	104.6	19.08
0.6	2.87	108.0	18.17

4.3.2.1 Surfactant Coverage Effects

To test the effects of surfactant coverage c_{in} on the droplet deformation under simple shear flow, the surfactant elasticity was set to $E_0 = 0.2$ as the use of this value was justified by Velankar et al. (2002) for low-molecular-weight surfactants. The saturation surfactant concentration was calculated using Eq. (4.14) and the resulting value was $\Gamma_\infty = 1.2 \times 10^{-4}[lmol/lu^2]$. This allowed the selection of the various initial surfactant concentrations Γ_i in order to achieve the range of surfactant coverage $0.2 \leq c_{in} \leq 0.6$. The surface Pèclet number was set to $Pe_s = 10$.

The interfacial area generation was calculated for the various cases at a dimensionless time step $\dot{\gamma}t = 1.17$ corresponding to the greatest value for the ratio $\frac{R_3}{R_0}$ which was presented in Fig. 4.7. The results shown in Fig. 4.8 indicate an increase in the area generation with the increase in the surfactant coverage as a consequence of the simultaneous increase in the droplet elongation (R_1), and flattening (R_3).

The dimensionless length R_1/R_0, the percentage elongation increase relative to the clean drop, and the reference angle $\theta°$ of the droplet inclination with respect to the horizontal direction were calculated at dimensionless time step $\dot{\gamma}t = 3.12$. The results presented in Table 4.1 imply that the greater the surfactant coverage the

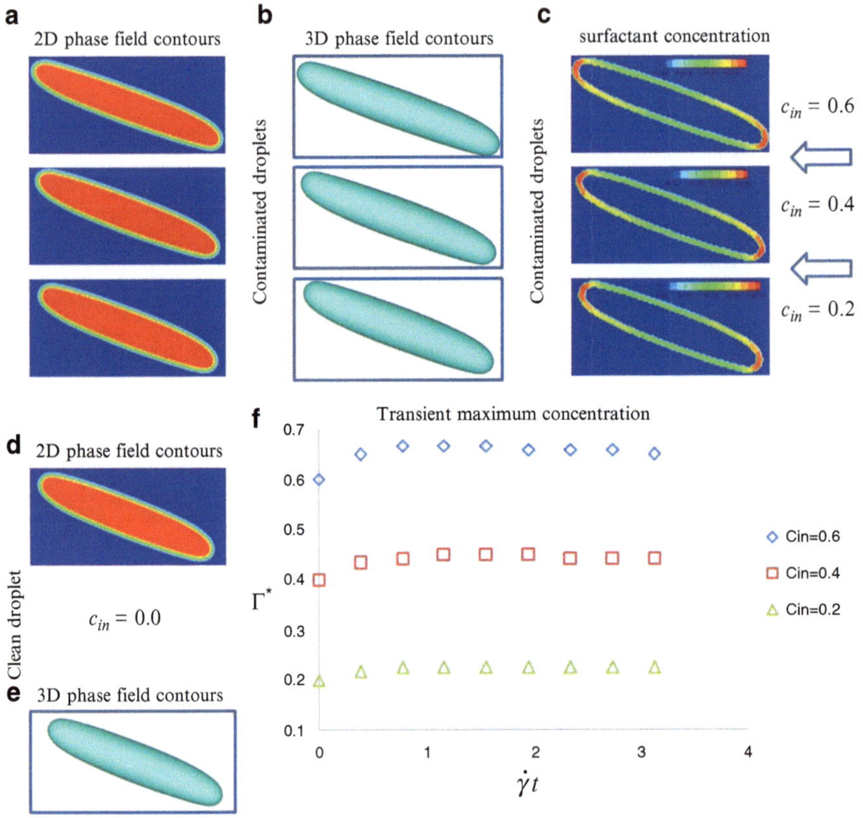

Fig. 4.9 (**a**) 2D view of the phase field contours from the central xz plane for the contaminated droplets. (**b**) 3D view of the phase field contours for the contaminated droplets surrounded by a fictitious block to show the variance in their dimensions. (**c**) 2D xz plane view of the surfactant concentration contours. (**d**) 2D xz plane view of the phase field contour for a clean drop. (**e**) 3D view of the phase field contour for the clean drop. (**f**) Graph representing the transient maximum values of the dimensionless surfactant concentration relative to dimensionless times. The results are for dimensionless time step $\dot{\gamma}t = 3.12$, capillary number $Ca = 4.6$, surfactant elasticity $E_0 = 0.2$, and Pèclet number $Pe_s = 10$

higher the values of the dimensionless length, the percentage elongation, and the lower the angle of the droplet inclination.

It is clear from the surfactant concentration contours in Fig. 4.9c (see Movie C4.3.2.1—Surfshear) that the regions of higher surfactant concentration are located around the tips of the droplet (see phase field density contours Fig. 4.9a and Movie C4.3.2.1—3Dshear) in the directions of the walls velocities, as a consequence of the convection of surfactants on the droplet interface. This also led to a greater droplet deformation as this was evident from the results of Table 4.1.

The phase field contours for clean and contaminated droplets and the surfactant concentration contours corresponding to the various values of the surfactant coverage for dimensionless time $\dot{\gamma}t = 3.12$ are presented in Fig. 4.9a–e.

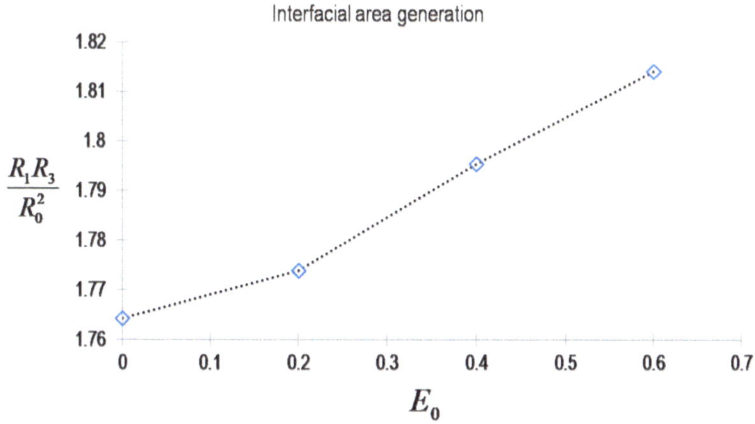

Fig. 4.10 Interfacial area generation for droplets in simple shear flow with respect to surfactant elasticity presented at a dimensionless time step $\dot{\gamma}t = 1.17$ and capillary number $Ca = 4.6$

The graph in Fig. 4.9f shows the transient values of the maximum dimensionless surfactant concentration with respect to the dimensionless time. The graph indicates a temporal increase in the maximum concentration due to convection followed by a slight decrease in these values, which is an expression on the interplay between the convection and the dilution of the surfactant due to the increase in the interfacial area of the droplet. This effect is slightly less pronounced in the case of surfactant coverage $c_{in} = 0.2$ since the droplet surface is relatively smaller.

4.3.2.2 Surfactant Elasticity Effects

Surfactant elasticity constants in the range of $0.2 \leq E_0 \leq 0.6$ were used to test the effects of the elasticity on the deformation of the droplet in simple shear flow. The saturation surfactant concentration was modified to affect a change in the surfactant elasticity since the presented model was isothermal, and several values were calculated by Eq. (4.14) which gave the following range of saturation concentration $1.2 \times 10^{-4} \leq \Gamma_\infty \leq 3.6 \times 10^{-4}[lmol/lu^2]$. The values for the initial concentration Γ_i were selected accordingly in order to maintain constant surfactant coverage $c_{in} = 0.2$. A surface Pèclet number $Pe_s = 100$ was used in the simulations. The interfacial area generation was calculated for the various cases with varying surfactant elasticity at the dimensionless time step $\dot{\gamma}t = 1.17$. The results presented in Fig. 4.10 shows an increase in the area generation with the increase in the surfactant elasticity.

The dimensionless droplet length, percentage elongation, and the angle of inclination of the droplet relative to the horizontal axis are presented in Table 4.2. A trend of increased dimensionless length, percentage elongation, and decreased

Table 4.2 Transient dimensionless length, percentage elongation, and angle of inclination measured in degrees, for three initial values of surfactant elasticity E_0 at dimensionless time step $\dot{\gamma}t = 3.12$

E_0	R_1/R_0	%	θ°
0	2.66	100.0	20.3
0.2	2.72	102.1	19.45
0.4	2.76	103.8	19.18
0.6	2.82	106.0	19.06

inclination angle of the droplet with the increase of the surfactant elasticity is observed from the results of Table 4.2.

The angle of inclination is dependent on the Marangoni stress which was given by Eq. (4.21). At the droplet caps the interfacial tension changes largely; hence larger Marangoni stress forces the droplet to align with the flow direction (Drumright-Clarke and Renardy 2004). The slow decrease in the inclination angle between the three contaminated droplets cases is indicative of the small difference in their Marangoni stress due to the constant surfactant coverage used in the various cases and the similarity of the flow conditions.

The phase field contours for clean and contaminated droplets, and the surfactant concentration contours are shown in Fig. 4.11a–e. Lower regions of surfactants concentration are observed around the droplet waist and higher regions around its tips. The graph in Fig. 4.11f shows the droplet transient minimum dimensionless surfactant concentration relative to the dimensionless time steps.

The minimum concentration decreased steadily due to the combined effects of surfactant convection and dilution. Slightly higher values were observed in the case of $E_0 = 0.2$ due to a relatively smaller droplet interfacial surface.

4.3.2.3 Pèclet Number Effects

The effects of the Pèclet number were investigated for the simple shear flow case with surfactant elasticity $E_0 = 0.2$, surfactant coverage $c_{in} = 0.2$, and saturation concentration $\Gamma_\infty = 1.2 \times 10^{-4} [lmol/lu^2]$. The following range of Pèclet numbers was used $1 \leq Pe_s \leq 100$.

The change in the interfacial area generation due to the change in the surface Pèclet number is presented in Fig. 4.12. The graph shows very minimal decrease in the area generation with the increase in the values of the Pèclet number. This could be explained by the fact that at low Pèclet numbers, surfactants are resistant to convection. Therefore in greater part of the interface, the droplet interfacial tension is lower than that of the clean droplet. This increases the droplet width R_3 and helps increase the area generation.

The dimensionless length, the percentage elongation, and the angle of the droplet inclination are presented in Table 4.3. The results reveal a marginal Pèclet number influence on the droplet elongation, because at low surfactant coverage, and

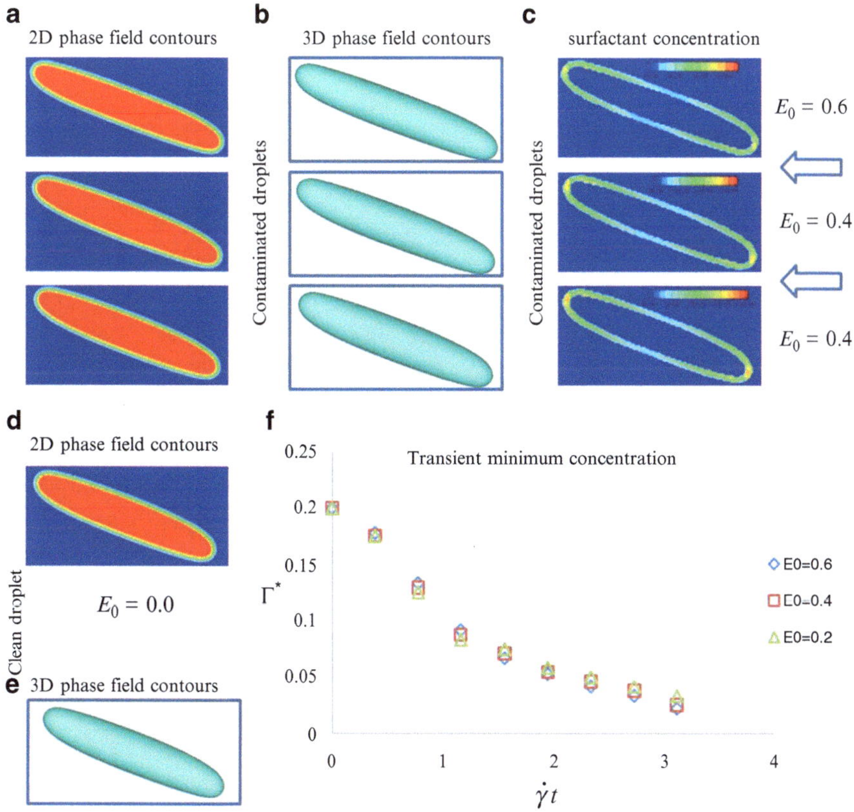

Fig. 4.11 (**a**) 2D view of the phase field contours from the central xz plane for clean droplets. (**b**) 3D view of the phase field contours for the contaminated droplets. (**c**) 2D view of the surfactant concentration contours in the xz plane. (**d**) 2D view of the phase field contour for the clean droplet in the xz plane. (**e**) 3D phase field contour for the clean droplet. (**f**) The graph represents the transient minimum values of the dimensionless surfactant concentration relative to dimensionless times. The results are for dimensionless time step $\dot{\gamma}t = 3.12$, capillary number $Ca = 4.6$, surfactant coverage $c_{in} = 0.2$, and Pèclet number $Pe_s = 100$

low surfactant elasticity, severe flow conditions are required to increase the surfactant concentration to levels capable of affecting a substantial influence on the interfacial tension due to the nonlinear nature of the surfactant equation of state used in this work.

The dimensionless surfactant concentration values Γ^* were calculated in a central xz plane along the lower circumference of the droplet. The calculation was done in the direction of the major axis. The coordinates $x_{m,a} = x/\cos\theta$ were normalized by the droplet radius R_0, and the corresponding dimensionless concentrations are shown in Fig. 4.13d. The phase field and surfactant concentration contours are shown in Fig. 4.13a–c.

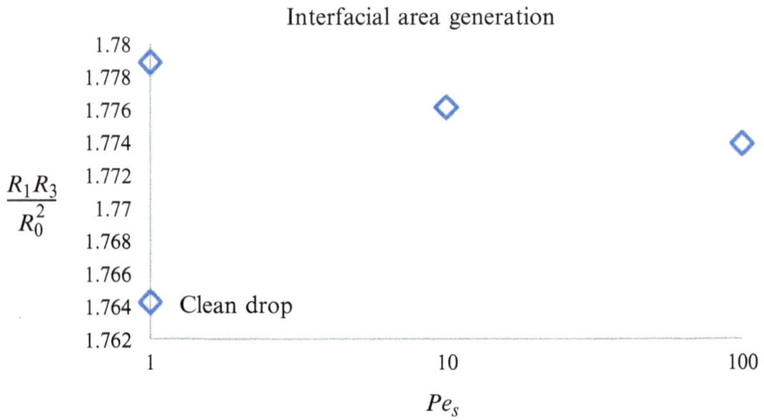

Fig. 4.12 Interfacial area generation for droplets in simple shear flow with respect to the surface Pèclet number presented at a dimensionless time step $\dot{\gamma}t = 1.17$ and capillary number $Ca = 4.6$

Table 4.3 Droplet transient dimensionless length, percentage elongation, and angle of inclination measured in degrees for $1 \leq Pe_s \leq 100$ at dimensionless time step $\dot{\gamma}t = 3.12$

Pe_s	R_1/R_0	%	$\theta°$
0	2.69	101.3	19.74
10	2.71	101.9	19.67
100	2.72	102.1	19.45

It is evident from Fig. 4.13c, d that at high Pèclet numbers more surfactants are convected towards the tips and away from the middle of the droplet, and this is due to the dominance of the convection role over the diffusion effects.

4.3.3 *Surfactant-Covered Droplets in Uniaxial Extensional Flows*

To simulate uniaxial extensional flow in the same domain used for the previous cases, the velocity field was modified as follows:

$$\boldsymbol{u}^{\infty}(\boldsymbol{x}) = \dot{\gamma}\begin{pmatrix} 1 & 0 & 0 \\ 0 & -0.5 & 0 \\ 0 & 0 & -0.5 \end{pmatrix} \cdot \boldsymbol{x} \tag{4.25}$$

where shear rate $\dot{\gamma} = 1.68 \times 10^{-5}\left[ts^{-1}\right]$ was used, yielding a capillary number $Ca = 0.8$.

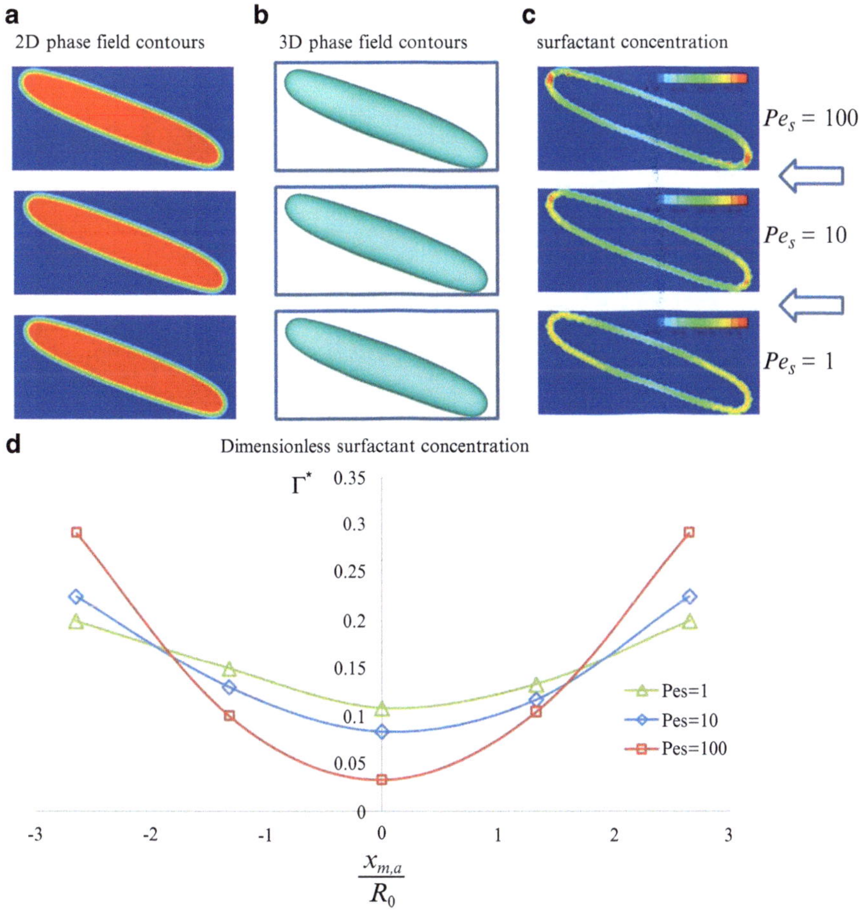

Fig. 4.13 (**a**) 2D xz view of the droplets phase field contours. (**b**) 3D view of the droplet phase field contours. (**c**) 2D view of the surfactant concentration contours. (**d**) The graph represents the dimensionless surfactant concentration relative to a position measured on the lower droplet circumference in the direction of the major axis. The results are for dimensionless time step $\dot{\gamma}t = 3.12$, surfactant elasticity $E_0 = 0.2$, surfactant coverage $c_{in} = 0.2$, and capillary number $Ca = 4.6$

4.3.3.1 Surfactant Coverage Effects

The surface Pèclet number was set to $Pe_s = 100$ and the surfactant elasticity to $E_0 = 0.2$. The surfactant coverage was varied in the range of $0.2 \leq c_{in} \leq 0.6$. All other fluid properties were kept unchanged. The simulation domain for a central droplet subjected to uniaxial extensional flow is shown in Fig. 4.14.

The surfactant concentration distribution under such flow conditions is symmetric, and it exhibits higher values towards the tips of the droplet and lower values in the center as this is shown in Fig. 4.15a–c (see Movies C4.3.3.1—3Dextensional

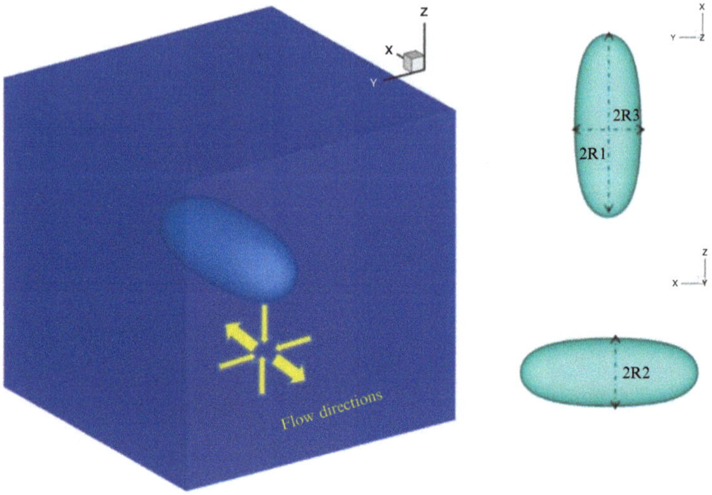

Fig. 4.14 Phase field contours for a central droplet in uniaxial extensional flow

and C4.3.3.1—Surf_ext). The graph in Fig. 4.15d represents the dimensionless surfactant concentration with respect to the x coordinates normalized by the droplet radius R_0.

It is clear from the 3D phase field contours insets of Fig. 4.15 that the droplet dimensionless length increased with the increase of the surfactant coverage.

4.3.3.2 Capillary Number Effects

To check the effect of the capillary number on the transient droplet behavior in extensional flows the following range of numbers was used $0.8 \leq Ca \leq 1.2$.

The dimensionless length R_1/R_0, the maximum dimensionless surfactant concentration Γ^*_{max}, and the minimum dimensionless surfactant concentration Γ^*_{min} were calculated at the dimensionless time step $\dot{\gamma}t = 0.604$, which corresponded to the end of the simulation time. It is clear from the presented results in Fig. 4.16d that the effects of surfactant dilution are dominant under supercritical capillary numbers, since the dimensionless length increased and both maximum and minimum surfactant concentration decreased with the increase in the capillary numbers.

No lamellar structure ($R_2 = R_3$) was observed under the uniaxial extensional flow conditions, since the compressional components in the xy and the xz planes were equivalent contrary to the case of the simple shear flow in which the compressional components were unbalanced. This is also true because the lamellar morphology is flow and viscosity ratio dependent for cases with finite capillary numbers (Cristini et al. 2002). Another factor which could potentially help the formation of the lamellar structure in simple shear flows is due to the nature of surfactant concentration distribution on the peripheries of a contaminated droplet in the low interface tangential velocity regions of the flow in the yz planes as shown in

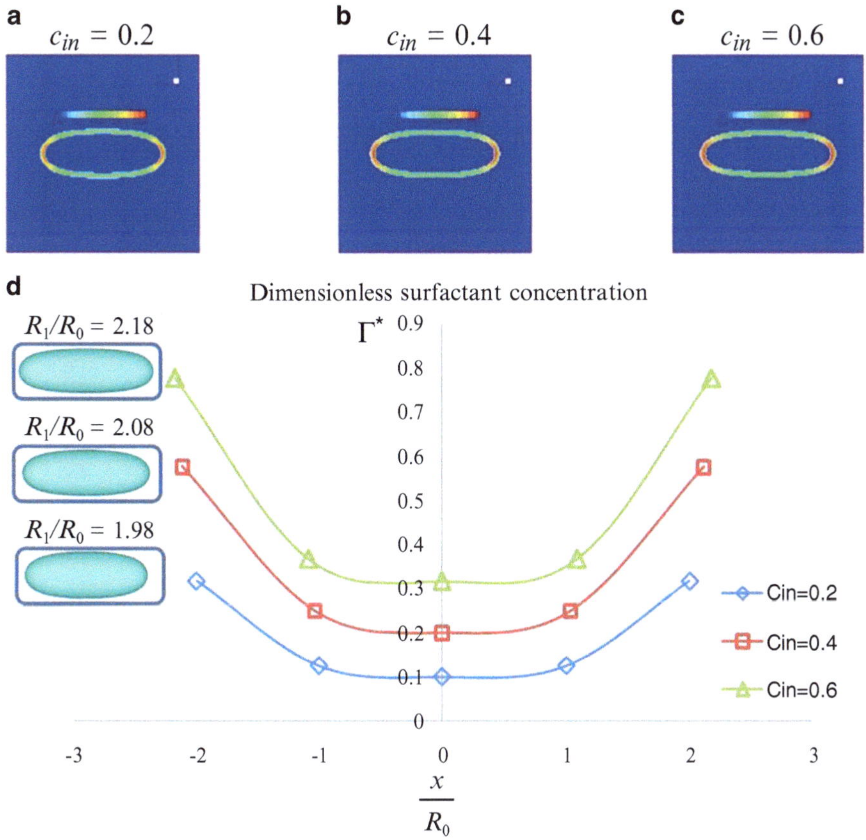

Fig. 4.15 (**a–c**) 2D xz view of the surfactant concentration contours for a central droplet in uniaxial extensional flow, for three values of the surfactant coverage. (**d**) Graph representing the dimensionless surfactant concentration in the xz plane as a function of the horizontal coordinate normalized by the droplet radius for $\dot{\gamma}t = 0.604$, $Ca = 0.8$, $Pe_s = 100$, and $E_0 = 0.2$. The *insets* in the graph are for the 3D view of the phase field contours

Fig. 4.17b. These regions are characterized by lower convection effects leading to higher local surfactant concentrations which act to reduce the droplet interfacial tension, hence locally lowering its capillary number and making it more deformable. This does not occur in the uniaxial extensional flow due to its uniform tangential velocity profile in the indicated region of Fig. 4.17a.

4.3.4 *Buoyancy of Surfactant-Covered Droplets in Infinite Medium*

The effect of surfactants on buoyant droplets and bubbles named here as fluid particles was studied both experimentally (Almatroushi and Borhan 2004; Griffith 1962; Bel Fdhila and Duineveld 1996; Alves et al. 2005) and numerically

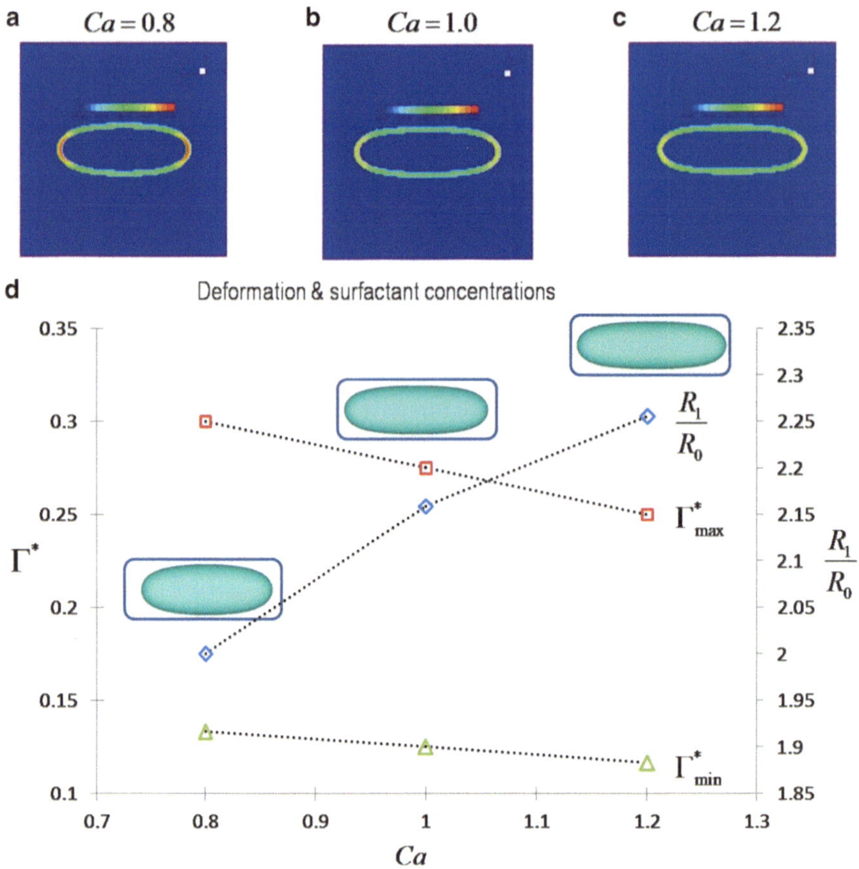

Fig. 4.16 (**a–c**) 2D view of the surfactant concentration contours on droplets in uniaxial extensional flow for a range of capillary number $08 \leq Ca \leq 1.2$. (**d**) Graph representing the values of the droplet dimensionless R_1/R_0, the dimensionless maximum $\Gamma^*_{\max}$, and minimum $\Gamma^*_{\min}$ surfactant concentration, respectively at dimensionless time step $\dot{\gamma}t = 0.604$. The *insets* in the graph are for the 3D view phase field contours

(Bel Fdhila and Duineveld 1996; Tasoglu et al. 2008). It was found that surfactants generally reduce significantly the particle's terminal velocity below the classical Hadamard-Rybszynski prediction in the spherical region of the shape regime; however in other shape regions the particle retardation due to surfactants is less effective (Tasoglu et al. 2008).

Buoyancy-driven fluid particles are characterized by the following dimensionless numbers:

$$E_o = \frac{g\Delta\rho d^2}{\sigma_0}, M_o = \frac{g\mu_f^4\Delta\rho}{\rho_f^2{\sigma_0}^3}, \mathrm{Re} = \frac{U_T d\rho_f}{\mu_f} \tag{4.26}$$

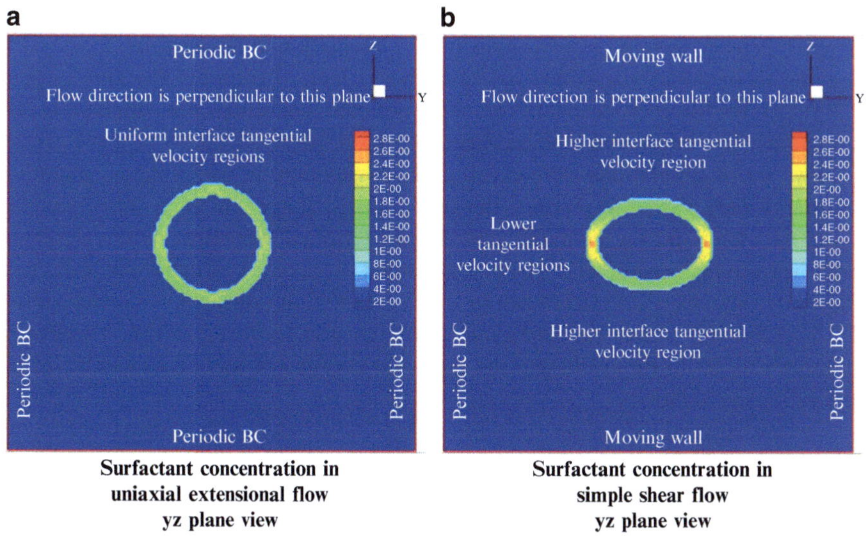

Fig. 4.17 (**a**) 2D yz view of the surfactant concentration contours is shown for a central droplet in uniaxial extensional flow. (**b**) 2D yz view of the surfactant concentration of a droplet in simple shear flows. The capillary number for the extensional flow is $Ca = 0.8$ and the dimensionless time step is $\dot{\gamma}t = 0.536$. The capillary number for the shear flow is $Ca = 4.6$ and the time step is $\dot{\gamma}t = 1.95$

where E_o is the Eotvos number, M_o is the Morton number, and Re is the Reynolds number, d is the particle diameter, σ_0 is the interfacial tension, g is the acceleration constant, μ_f is the ambient fluid dynamic viscosity, and U_T is the particle terminal velocity.

The predicted Hadamard-Rybszynski terminal velocity for a spherical fluid particle rising in infinite medium is given by Clift et al. (2005):

$$U_{HR} = \frac{2}{3}\frac{ga^2\Delta\rho}{\mu_f}\frac{\mu_f + \mu_d}{2\mu_f + 3\mu_d} \tag{4.27}$$

where a is the particle radius and μ_d is the particle dynamic viscosity.

The terminal velocity for a rising fluid particle in infinite medium with $E_0 > 40$ and $M_0 > 200$ can be estimated analytically through solving the following equation (Clift et al. 2005):

$$2\text{Re}^2 + 6\text{Re}\frac{2 + 3\eta}{1 + \eta} - E_o^{3/2}M_o^{-1/2} = 0 \tag{4.28}$$

where $\eta = \mu_d/\mu_f$ is the viscosity ratio.

A domain consisting of $61 \times 61 \times 351[lu^3]$ was used to investigate the effects of surfactants on the terminal velocity of a single droplet of radius $a = 12[lu]$, rising in an infinite medium, in the spherical and the spherical cap regions, respectively. The density of the suspending fluid was $\rho_f = 2[mu/lu^3]$, and the droplet density was $\rho_d = 1.0[mu/lu^3]$. The relaxation time was set to $\tau = 1.0$ leading to dynamic viscosities $\mu_f = 0.333$ and $\mu_d = 0.1666$ for the surrounding fluid and the droplet, respectively. The periodic condition was applied in all directions. The interfacial tension was set to $\sigma_0 = 10^{-3}$.

Two set of simulations were executed in this section, in which the acceleration constants were varied to produce the two required shapes. In each set of simulations the clean and the contaminated droplet terminal velocities were calculated and measured consecutively. The contaminated droplets cases had their surfactant elasticity set to $E_0 = 0.5$ leading to a saturation surfactant concentration $\Gamma_\infty = 1.5 \times 10^{-3}$ and the surfactant coverage was set to $c_{in} = 0.4$. The Pèclet number was set to $Pe_s = 40$ and the diffusion coefficient was evaluated based on the following relation:

$$Pe_s = \frac{U_T \ell}{D_s} \tag{4.29}$$

where U_T is the terminal velocity, ℓ is the vertical distance required for the clean droplet to attain its terminal velocity.

The resulting terminal velocities of the clean and the contaminated buoyant droplets in the spherical shape region were due to acceleration constant $g = 2.0 \times 10^{-5}[lu/ts^{-2}]$ which led to Morton number $M_o = 61.7$, Eotvos number $E_o = 11.5$, and Reynolds number Re $= 0.34$. The normalized terminal velocities by the Hadamard-Rybszynski terminal velocity U_{HR} and the normalized mass center vertical locations by the domain length L calculated with respect to normalized time by $\sqrt{d/g}$ are shown in Fig. 4.18.

The terminal velocities of the clean and contaminated buoyant droplets in the spherical-cap shape region were produced by acceleration constant $g = 2.0 \times 10^{-4}[lu/ts^{-2}]$ which yielded Morton number $M_o = 617$, Eotvos number $E_o = 115$, and Reynolds number Re $= 2.6$. The normalized terminal velocities by the theoretical droplet terminal velocity U_{Th} and the normalized mass center vertical locations by the domain length L calculated with respect to normalized time by $\sqrt{d/g}$ are shown in Fig. 4.19.

A third simulation was executed using the parameters from the second case with $g = 4.0 \times 10^{-4}[lu/ts^{-2}]$ and the results for the phase field, the surfactant concentration distribution, and velocity contours are shown in Fig. 4.20. The increase in the acceleration resulted in Morton number $Mo = 1234$, Eotvos number $Eo = 230$, and Reynolds number Re $= 4.32$. Under such condition higher droplet deformation and very marginal effects of surfactant retardation are witnessed.

The results for the terminal velocities and their ratio with respect to the theoretical values calculated by Eqs. (4.27) and (4.28) from both set of simulations are summarized in Table 4.4.

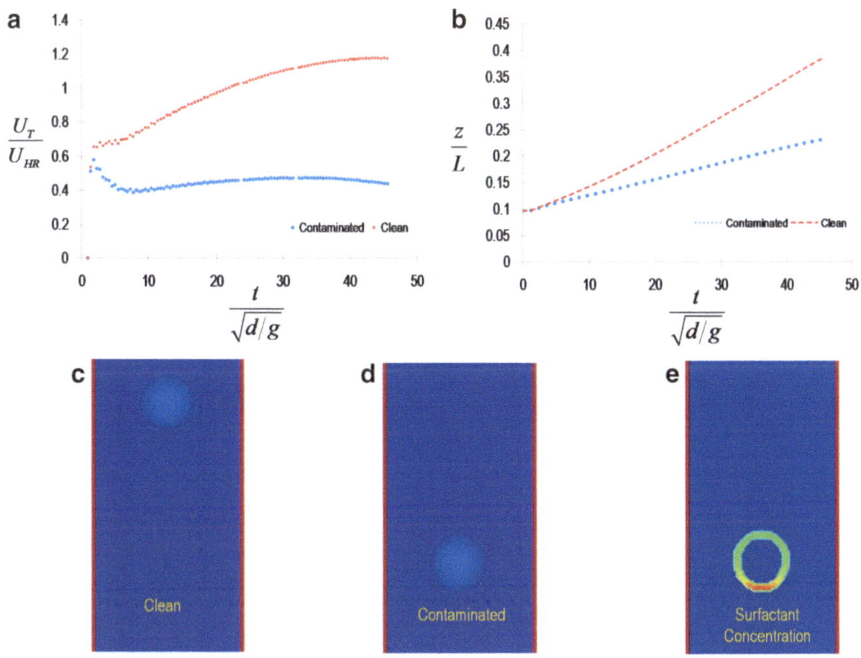

Fig. 4.18 (**a**) Dimensionless terminal velocities for clean and contaminated buoyant droplets presented in the spherical region. (**b**) Dimensionless vertical location of the droplet mass center for the clean and the contaminated droplets. (**c**) 3D phase field contours for the clean droplet, (**d**) 3D phase field contour for the contaminated droplet. (**e**) 2D xz plane view of the surfactant concentration contour for the contaminated droplet. The dimensionless time step is $t/\sqrt{d/g} = 43.3$. The simulation was executed with acceleration constant $g = 2.0 \times 10^{-5}$, Morton number $M_o = 61.7$, Eotvos number $E_o = 11.5$, and Reynolds number $\mathrm{Re} = 0.34$

The surfactant concentration phase field contours in Figs. 4.18e, 4.19e, and 4.20c and the terminal velocity results from Table 4.4 are clear indications of the presented model compliance with the known physical mechanism governing the buoyant droplet behavior due to the influence of surfactants which was introduced by Frumkin and Levich (Tasoglu et al. 2008). This mechanism considers the droplet retardation as a result of the surfactants convection towards the back of the droplet, which in turn creates Marangoni stress and slows down its surface mobility. This leads to increased drag force and decreased terminal velocity. The model also shows that the effects of surfactants on the droplet terminal velocity are more influential in the spherical region relative to the spherical-cap region.

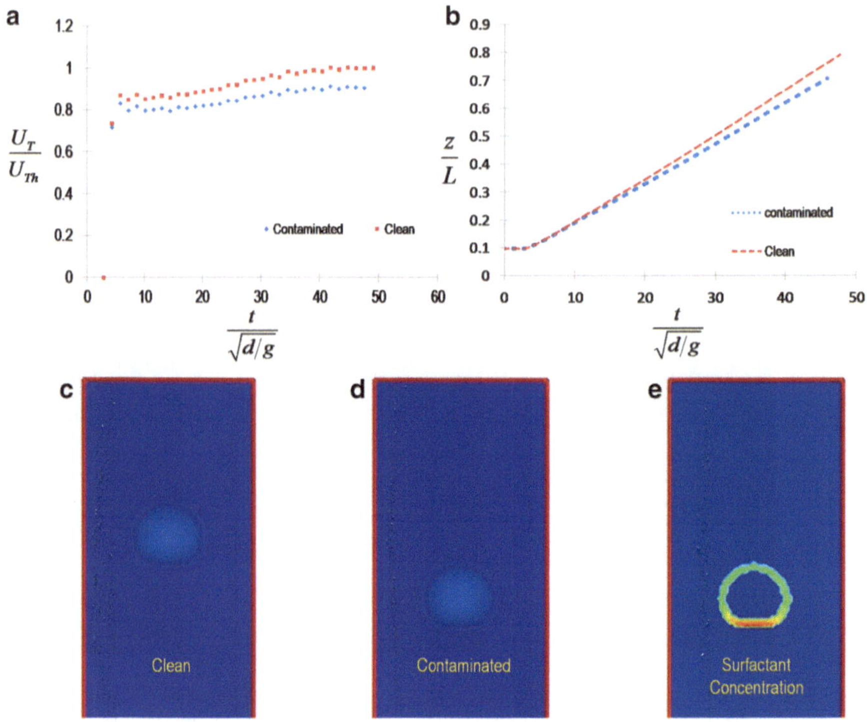

Fig. 4.19 (**a**) Dimensionless terminal velocities for clean and contaminated buoyant droplets presented in the spherical-cap region. (**b**) Dimensionless vertical location of the droplet mass center for the clean and the contaminated droplets. (**c**) 3D phase field contours for the clean droplet. (**d**) 3D phase field contour for the contaminated droplet. (**e**) 2D view of the surfactant concentration contour for the contaminated droplet. The dimensionless time step is $t/\sqrt{d/g} = 49.07$. The simulation was executed with acceleration constant $g = 2.0 \times 10^{-4}$, Morton number $M_o = 617$, Eotvos number $E_o = 115$, and Reynolds number $\mathrm{Re} = 2.6$

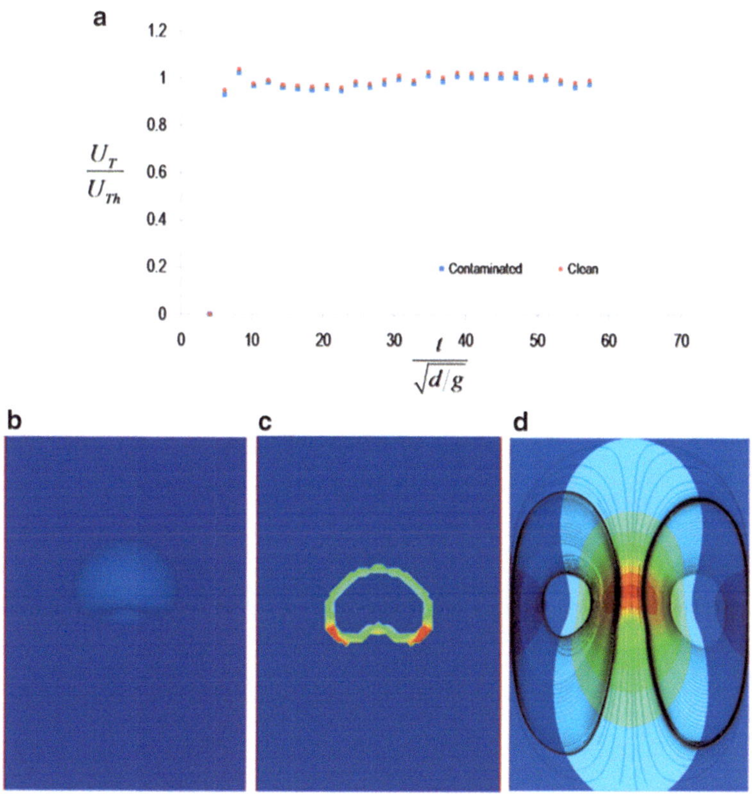

Fig. 4.20 (**a**) Dimensionless terminal velocities for clean and contaminated buoyant droplets in the spherical-cap region. (**b**) 3D slightly tilted view for the droplet phase field contour. (**c**) 2D view of the surfactant concentration contour for the contaminated droplet. (**d**) 2D view of the velocity contours in the vertical direction with some streamlines. The simulation was executed with acceleration constant $g = 4.0 \times 10^{-4}$, Morton number $Mo = 1234$, Eotvos number $Eo = 230$, and Reynolds number $\mathrm{Re} = 4.32$

Table 4.4 Comparison of the terminal velocities from the two sets of simulations in the spherical and spherical-cap regions with their respective theoretical terminal velocities

Condition	Region	U_r	$U_{HR/Th}$	%
Clean	Spherical	0.00290	0.00247	1.175
Contaminated	Spherical	0.0011	0.00247	0.451
Clean	Spherical-cap	0.01803	0.01801	1.001
Contaminated	Spherical-cap	0.01641	0.01801	0.911

Chapter 5
Suppressing the Coalescence in the LBM: Colloids Rheology

Abstract This chapter presents a novel phenomenological approach for suppressing the coalescence in the Gunstensen multicomponent lattice Boltzmann method (LBM). The suppression of coalescence is achieved by perturbing the terminal nodes of the ambient fluid's thin layer trapped between the approaching droplets. This additional perturbation creates a local high pressure fluid layer which eventually leads to suppressing the coalescence of the neighboring droplets while maintaining a suitable qualitative force balance representative of the physical intermolecular forces which act between them. The suppression of coalescence enables predicting the rheological properties of soft colloids, which is one of the most challenging and complicated problems in material and fluid sciences. Substantial complications in prediction of rheology arise due to the deformability and aggregation of soft colloids. Thus, a better understanding to deformation and aggregation of colloids can lead to a better understanding of the shear thinning region of colloidal solutions. This chapter provides a practical overview of the deformation and de-aggregation of droplets with change in time or increase in shear rate. Simulations performed here are used to quantify (a) relative viscosity of emulsions with change in shear rate, (b) relative viscosity of emulsions with change in time, (c) effect of deformation of droplets on the shear thinning region in emulsions, and (d) relative viscosity of emulsions with change in volume fraction.

The multicomponent lattice Boltzmann method (LBM) has emerged as a powerful tool for simulating a variety of fluid problems related to the droplet deformation, breakup, and coalescence. However this method is incapable of simulating cases where neighboring contaminated droplets of the same makeup amalgamate

---

Electronic supplementary material Supplementary material is available in the online version of this chapter at http://dx.doi.org/10.1007/978-1-4899-7402-0_5. Videos can also be accessed at http://www.springerimages.com/videos/978-1-4899-7401-3

H. Farhat et al., *Accelerated Lattice Boltzmann Model for Colloidal Suspensions: Rheology and Interface Morphology*, DOI 10.1007/978-1-4899-7402-0_5,

rather than coalesce. This problem can be solved in the LBM if different colors are assigned to the various droplets. The disadvantages of such an approach are the requirement for more computational resources (Dupin et al. 2003) and the undermining of the molecular interaction forces which act between the approaching droplets. The latter is important for determining the rheological behavior of colloids. Suppressing coalescence in the LBM enables the model to handle issues of aggregation and disaggregation under shear stress, and to deal with the rheology of polymers and colloids in a variety of flows under low and moderate Reynolds numbers. These problems are of fundamental and practical interest to many industries, since the rheology of such systems plays a decisive role in their transport properties, and physical and thermal qualities. This chapter presents a heuristic LBM scheme suitable for the study of emulsion rheology. The advantages of such a model are the simplicity in its implementation, the requirement of less computational resources, and most importantly is the breakaway from the empirical models, since the effective viscosity of the non-Newtonian fluids is calculated directly.

5.1 Suppressing the Coalescence in the LBM

Two droplets situated at a close proximity from each other in the multicomponent LBM tend to coalesce. The strength of the coalescence is dependent on the interfacial tension parameter α which determines the magnitude of the force in Eq. (2.16) responsible for the creation of a pressure difference between the neighboring droplets internal fluid and the narrow external fluid layer between them. Eventually the pressure difference leads to the depletion of the external layer and allows a direct contact between the droplets, which further destroys the droplets contacting interfaces and allows them to coalesce.

In the presented model, the LBM inherent coalescence driving force is used heuristically to represent the attraction forces between the approaching interfaces and a counteracting steric repulsion is employed to represent the repulsive forces. The following formula is proposed for indirectly imposing a repulsive force in the LBM with a magnitude:

$$S(\boldsymbol{x}) = c_{in}^{*}\lambda(\alpha)\alpha(\boldsymbol{x}) \tag{5.1}$$

where λ is a dimensionless function of the interfacial tension parameter, which is related to the local surfactant concentration through Eq. (4.20); $c_{in}^{*} = f(c_{in}/c_{in}^{\min})$ is a nonlinear function of the ratio of the initial surfactant coverage c_{in} and the minimum coverage $c_{in}^{\min}$ required for suppressing the coalescence in a particular mixture (Lyu et al. 2002). The application of the repulsive force is accomplished through

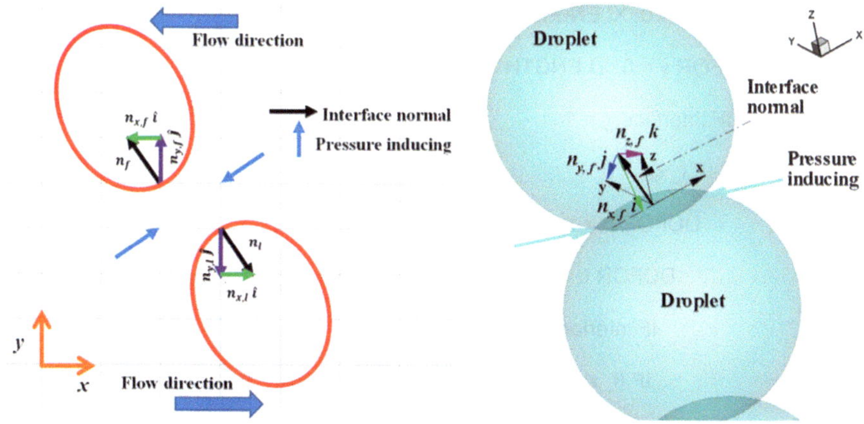

Fig. 5.1 Two approaching droplets in a hypothetical shear flow, with the required forces for suppressing the coalescence without altering the droplets shape

first identifying an interfacial node belonging to a leading drop in the computational sweeping direction. A forward small loop scans whether another interfacial node is within the vicinity of the leading node and located at a distance $(x_l - x_f)^2 + (y_l - y_f)^2 \leq 2$. The span of the loop depends on whether the leading droplet is approaching the following one in the horizontal or the vertical direction. For example for the horizontal direction, the loop is given by $x + 1 \leq p < x + 3$; $y - 2 \leq q < y + 3$, where x and y are the interface nodes coordinates in the main loop, p and q are the coordinate of the small loop. The normal to the interface components n_x, n_y signs are hence after checked. The fulfillment of the condition $\text{sgn}(n_{x,l}) \neq \text{sgn}(n_{x,f})$ and or $\text{sgn}(n_{y,l}) \neq \text{sgn}(n_{y.\,f})$ ensures that the two nodes belong to different droplets as shown in Fig. 5.1.

The following is pseudocode for identifying the nodes where the suppression of coalescence force should be applied

```
DOFOR x = 0, XLENGTH
 DOFOR y = 0, YLENGTH
  IF interface_x,y = 1
   IF f_x_x,y < 0
    DOFOR p = x+1, x+4
     DOFOR q = y-2, y+3
      IF interface_x,y = 1 AND (x-p)*(x-p) + (y-q)*(y-q) <= 2
       IF (f_x_x,y < 0 AND f_x_p,q > 0) OR (f_y_x,y > 0 AND f_y_p,q < 0)
        XMarker_x,y = 1
       END IF
      END IF
     END DO
    END DO
   END IF
  END IF
 END DO
END DO
```

*Similar implementation should be followed to identify nodes at which $\text{YMarker}_{x,y} = 1$

A set of forces are then applied in the tangential direction at the locations indicated by pressure inducing in Fig. 5.1 as follows:

$$\boldsymbol{F}(\boldsymbol{x}) = \pm S(\boldsymbol{x})\boldsymbol{T} \tag{5.2}$$

where $\boldsymbol{T}(n_y, -n_x)$ is the unit tangent to the leading droplet interface and $\boldsymbol{n}(n_x, n_y)$ is the unit normal vector. The tangential forces are only applicable to the ambient fluid; therefore only the two diagonally opposite forces shown in blue in Fig. 5.1 are active at any particular time.

The macroscopic force in Eq. (5.2) is applied through the source term of Eq. (2.11) by the following relation (Halliday et al. 2007):

$$\phi_i = \omega_i \frac{1}{k_2} \boldsymbol{F} \cdot \boldsymbol{c}_i \tag{5.3}$$

where the constant is given by $k_2 = 1/3$. The applied opposite body like forces act to create a pressure rise in the thin external fluid layer trapped between the two

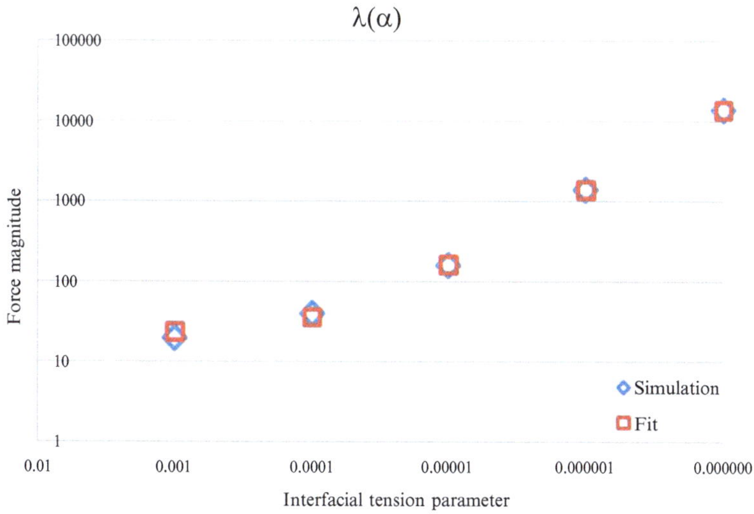

Fig. 5.2 The limiting condition for determining the dimensionless function $\lambda(\alpha)$ and its curve fit, for $c_{in}^{*} = 1$

droplets. The distributed pressure over the area of the local interfaces yields a repulsive force which prevents the droplets from coalescing, without causing any significant local droplets deformation.

5.1.1 *Optimizing the Force Equation for Suppressing Coalescence*

The repulsive force magnitude used in Eq. (5.1) includes a function $\lambda(\alpha)$, which was needed for controlling the force when the interfacial tension parameter varied locally without changing the initial surfactant coverage. Due to the surfactant concentration evolution on the droplet interface, $\lambda(\alpha)$ should affect an automatic change in the force magnitude corresponding to the changing local interfacial tension. The tension relation to the local surfactant concentration is governed by the Langmuir equation of state.

In line with the above reasoning several simulations with uniform interfacial tension were executed to optimize Eq. (5.1). A $123[lu^2]$ square lattice unit domain was used for investigating the required force magnitude needed for suppressing the coalescence of two droplets with radius $R = 18[lu]$ placed initially at a distance of $4[lu]$ between their interfaces in a simple shear flow. A shear rate of 9.75×10^{-5} was induced by the opposite movement of the upper and lower walls. Periodic condition was applied on the side boundaries. The objective was to find some optimal values for λ capable of effecting droplets total separation during couple thousand of time steps for the value $c_{in}^{*} = 1$. Three viscosity ratios μ_d/μ_s were utilized 0.1, 1, and 10, and the highest values for λ were taken. A graph was produced and curve fitted using an inverse first-order polynomial, to obtain the appropriate equation for $\lambda(\alpha)$. The graph is presented in Fig. 5.2.

Equation (5.1) was then replaced with the following formula:

$$S(\boldsymbol{x}) = c_{in}^{*}\left(22.2 + \frac{0.001398}{\alpha}\right)\alpha \tag{5.4}$$

by which for values $c_{in}^{*} < 1$ the force balance allows the occurrence of coalescence starting by a slow drainage from one droplet into the other for values closer to one, while the values $c_{in}^{*} \geq 1$ lead to suppressing the coalescence in the presented model.

The following is pseudocode for applying suppression of coalescence force at the identified nodes

```
DOFOR x = 0, XLENGTH
 DO FOR y = 0, YLENGTH
  IF XMarker_x,y = 1 AND interface_x,y = 1
   DOFOR I = 0, Q
    IF f_x_x,y < 0
     IF b_x-1,y[i] != 0
      b_x-1,y-1[i] = b_x-1,y-1[i] - 3*S(x)*σ_x,y*weighting_factor*(cx[i] * f_y_x,y- cy[i] * f_x_x,y)
     END IF
     IF b_x-1,y-2[i] != 0
      b_x-1,y-2 = b_x-1,y-2 - 3*S(x)*σ_x,y*weighting_factor*(cx[i] * f_y_x,y- cy[i] * f_x_x,y)
     END IF
     IF b_x+1,y-1[i] != 0
      b_x+1,y-1 = b_x+1,y-1 + 3*S(x)*σ_x,y*weighting_factor*(cx[i] * f_y_x,y- cy[i] * f_x_x,y)
     END IF
     IF b_x+1,y-2[i] != 0
      b_x+1,y-2 = b_x+1,y-2 + 3*S(x)*σ_x,y*weighting_factor*(cx[i] * f_y_x,y- cy[i] * f_x_x,y)
     END IF
    END IF
   END DO
  END IF
 END DO
END DO
```

*Similar force implementation should be followed when $\text{YMarker}_{x,y} = 1$

5.1.2 Test Cases

5.1.2.1 Suppressing the Coalescence in a Quiescent Flow

To demonstrate the capability of the proposed algorithm in suppressing the coalescence of several droplets in a quiescent flow, a domain consisting of $123[lu^2]$, with a central droplet of radius $R = 18[lu]$ and four satellite droplets of radii $R_s = 12[lu]$ positioned equidistantly from the domain center, was used. Periodic conditions were applied on the four boundaries. The interfacial tension parameter was $\alpha = 1.0 \times 10^{-4}$. The two fluids had the same density $\rho = 1.0[mu/lu^2]$ and the same kinematic viscosity $\nu = 0.166[lu^2ts^{-1}]$; hence the droplets were neutrally buoyant in their surrounding fluid. The selection of these conditions was dictated by a desire to maintain the droplets at a close proximity in the absence of any external forces, and monitor their behavior. Very small value of the Pèclet number was used to limit the convection of the surfactants on the interface of the droplets, which behaved as clean ones.

The phase field contours resulting from the model with the suppression of coalescence for $c^*_{in} = 1$ compared with the phase field contours of the standard model $c^*_{in} = 0$ are shown in Fig. 5.3 (see Movies C5.1.2.1—Coalescence, C5.1.2.1—Supcoal and C5.1.2.1—SupCoal4d) for several dimensionless time steps. A characteristic time step $\mu R/\sigma_0$ was used here where μ is the dynamic viscosity and σ_0 is the interfacial tension. It is evident from Fig. 5.3 that the coalescence was suppressed in the case of $c^*_{in} = 1$, contrary to the standard case

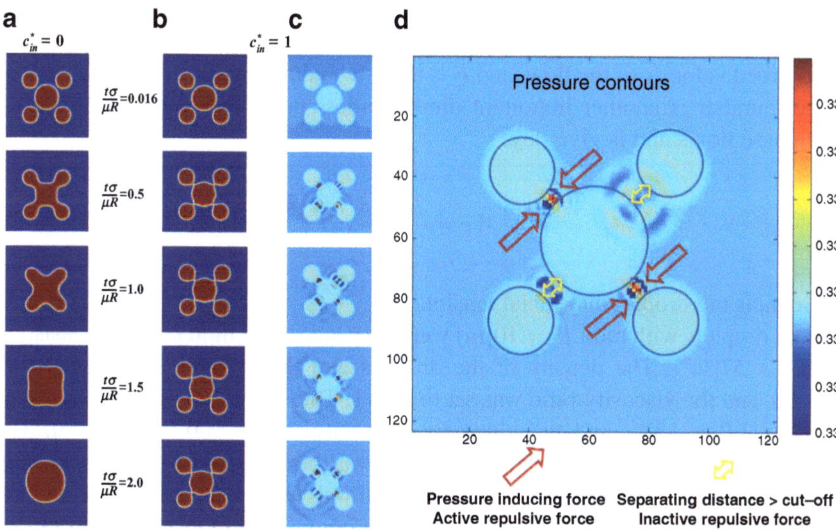

Fig. 5.3 (**a**) Phase field contours for droplets in quiescent flow and $c^*_{in} = 0$ for the various dimensionless time steps $t\sigma/\mu R$. (**b**) Phase field contours for the droplets with $c^*_{in} = 1$ for the same time steps. (**c, d**) Pressure contours for the droplets with time varying interaction forces

with $c_{in}^* = 0$ where a single drop resulted from the coalescence of all the droplets in the domain.

The pressure contours in Fig. 5.3 show the continuous attempt by the coalescence force to bring the droplets to a minimal proximity, counteracted by the induced repulsive force which helped creating a higher pressure film between the approaching interfaces, leading to droplet amalgamation instead of coalescence.

5.1.2.2 Suppressing the Coalescence in a Dynamic Flow

Parabolic flows are interesting in many research areas especially those related to droplet-based microfluidics. Another area of interest is the hemodynamics, since blood circulation in the various vessels is by nature parabolic. The red blood cells (RBC) behavior in the microvasculature could be studied through modeling the RBC as a deformable droplet. The disadvantage of such assumption in parabolic flows is related to the changing strain rate with respect to the droplet location in the domain. Hence droplets situated in the central region move at higher velocity than those moving closer to the walls. This diminishes the distance between them and eventually leads to their collision and coalescence, unless the coalescence is interrupted.

Multiphase parabolic flows are characterized by the channel Reynolds number:

$$\mathrm{Re}_{ch} = \frac{\overline{U}H}{\nu} \tag{5.5}$$

where H is the channel height, $\overline{U} = FH^2/12\rho\nu$ is the magnitude of the average undisturbed velocity of the flow, and $\boldsymbol{F}$ is the flow inducing macroscopic force. The Weber number is another important dimensionless parameter used for analyzing multiphase flows and is given by:

$$We = \frac{2\rho\overline{U}^2 R}{\sigma_0} \tag{5.6}$$

where σ_0 is the droplet interfacial tension.

Four droplets with radii $R = 10[lu]$ were suspended in a fluid domain consisting of $351 \times 57[lu^2]$. The density of the droplets and the ambient fluid was set to $\rho = 1.0$, and the viscosity ratio was set to $\eta = 1$. The interfacial tension parameter was $\alpha = 1.0 \times 10^{-3}$, and the kinematic viscosity was $\nu = 0.166[lu^2ts^{-1}]$. The bounce back was applied on the upper and lower boundaries to impose a no-slip velocity condition on the walls. Periodic condition was used for the inlet and the outlet of the domain, respectively. A force $F = 2.5 \times 10^{-6}[mulu/ts^{-2}]$ was applied in the horizontal positive direction, and this led to a Reynolds number $\mathrm{Re}_{ch} = 1.4$ and Weber number $We = 0.33$. The characteristic time was taken as the inverse of the strain rate $\dot{\gamma}_{H/4}^{-1}$ calculated for a distance equivalent to quarter of the

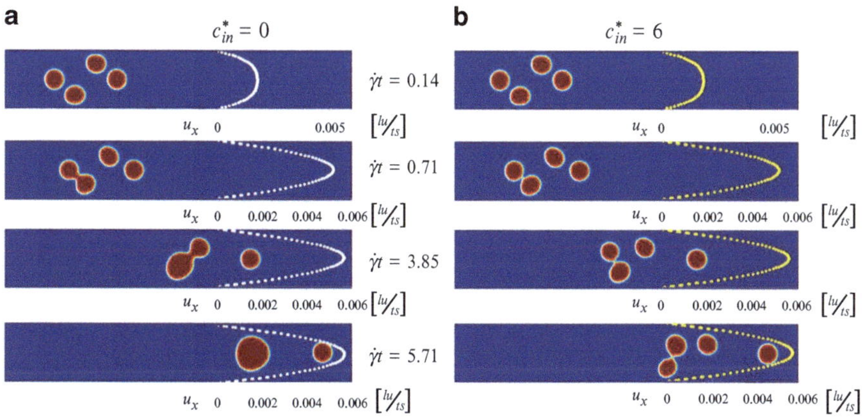

Fig. 5.4 (**a**) Phase field contours from the standard LBM $c^*_{in} = 0$, in which coalescence occurs instantaneously after collision. (**b**) Phase field contours from the presented model $c^*_{in} = 6$ in which the coalescence is suppressed. The horizontal velocity profile at $x/H = 0.77$ is superimposed on the phase field contours

channel height. The phase field contours from the standard LBM $c^*_{in} - 0$ and the presented model $c^*_{in} = 6$ are shown in Fig. 5.4.

It is obvious from Fig. 5.4b that the coalescence was suppressed in the presented model, and this has revealed the following observations. The collision between the two droplets was due to the higher velocity of the central drop. This led to a change in the trajectories and the velocities of the collided droplets. The first droplet velocity decreased while it was riding over the second droplet as this was evident from the increasing dimensionless distance between the mass centers of the first and the fourth droplets as the simulation time evolved $(d/H)_{\dot{\gamma}t=0} = 1.04$ versus $(d/H)_{\dot{\gamma}t=5.7} = 1.61$. The second droplet shear lift was partially undermined, since its mass center dimensionless distance from the lower wall $(y_{lo}/H)_{\dot{\gamma}t=5.7} = 0.26$ was smaller than the one between the third droplet and the upper wall $\left(H - y_{up}/H\right)_{\dot{\gamma}t=5.7} = 0.35$ at the end of the simulation time.

5.2 Rheology

5.2.1 *Direct Calculation of the Relative Viscosity of Colloids*

Coaxial viscometers are used for determining the viscosity of fluids by measuring the torque needed to keep one of their cylinders stationary while the other rotates with the fluid placed between them. The torque induced on the stationary cylinder is proportional to the effective viscosity of the tested fluid, while the shear rate is decided by the rotational speed of the other. Following the same principle of the

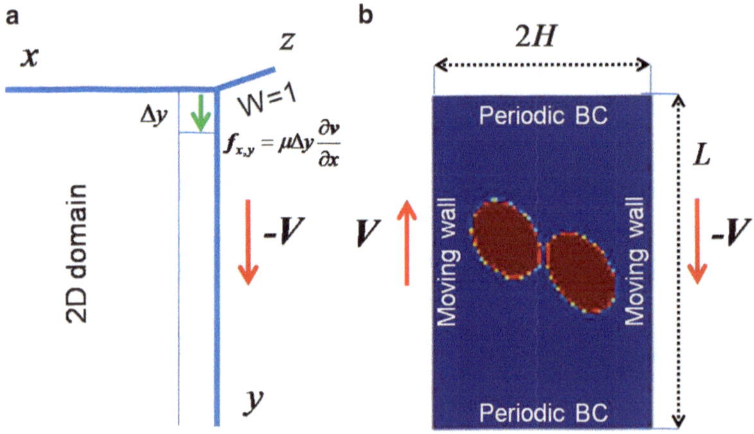

Fig. 5.5 (**a**) Schematic of the 2D domain of thickness $W = 1$ used for the derivation of the effective viscosity. (**b**) Phase field contours for two droplets in a simple shear flow with an indication of the used boundary conditions

coaxial viscometers operation, the schematic in Fig. 5.5 shows the proposed model's domain used for the derivation of the effective viscosity, and the phase field of two noncoalescing droplets due to the application of the suppression of coalescence algorithm.

Assuming that μ is the suspending fluid dynamic viscosity, V is the linear velocity of the walls, H is half the distance between the walls in the horizontal direction, and L is the length of the walls in the vertical direction, the shearing force per unit width exerted by the ambient fluid on the moving wall without suspended immiscible droplets is $|\boldsymbol{F}| = \mu \frac{|V|}{H} A$. The undisturbed flow velocity in the y direction is $v = \frac{|V| x}{H}$. With suspended droplets the effective viscosity of the whole fluid is $\mu_{eff} = \frac{H|F|}{L|V|}$. The thickness in the z direction is assumed as unity, thus the area $A = L$ and the total shearing force is:

$$|\boldsymbol{F}| = \sum |f_{x,y}| = \mu \Delta y \Delta z \sum \frac{\partial v}{\partial x} \tag{5.7}$$

This leads to the following formula for the effective viscosity:

$$\mu_{eff} = \mu \frac{H}{|V|} \frac{\Delta y \Delta z}{L} \sum \frac{\partial v}{\partial x} \tag{5.8}$$

The relative viscosity is the ratio of the effective viscosity and the viscosity of the suspending fluid, and it is given for an equidistant grid with spacing $\Delta y = \Delta x = 1$ by:

$$\mu_{rel} = \frac{H}{|V|L} \sum \frac{\partial v}{\partial x} \tag{5.9}$$

From Eq. (5.8) it is obvious that the effective viscosity is mainly decided by the change in the local shear rate near the wall, which is the only variable both spatially and temporally.

5.3 Simulation Results and Discussions

5.3.1 Simulation Method

Various cases simulated to study the viscosity of emulsions at different shear rates and volume fractions are tabulated and shown in Table 5.1. Two different droplet diameters of 13.33 and 20 μm were used in 3D simulations. When the droplet size was 20 μm, computational domain considered for 3D study is a cubical box with dimensions of 100 × 30 × 100 μm; whereas the dimensions are 100 × 100 μm for 2D study. When the size of droplet was taken as 13.33 μm, the computational domain had dimensions of 100 × 18 × 100 μm. Viscosity ratio between the droplets and continuous fluid is considered as $\lambda = 7.26$. Shear force was applied

Table 5.1 Various simulations performed to study the relative viscosity of emulsions are shown in the Table. 3D simulations with droplet diameter of 13.33 μm were only performed for volume fraction of $\phi = 0.29$

Simulation method	Shear rate (s^{-1})	Droplet diameter (μm)	Range of volume fraction (ϕ)	Total time steps
2D	0.06	20	0.15, 0.30, 0.38	5,000,000
2D	0.2	20	0.15, 0.30, 0.38	1,500,000
2D	0.6	20	0.15, 0.30, 0.38	500,000
2D	1.8	20	0.15, 0.30, 0.38	170,000
2D	3.0	20	0.15, 0.30, 0.38	100,000
2D	12.0	20	0.15, 0.30, 0.38	25,000
2D	24.0	20	0.15, 0.30, 0.38	12,500
3D	0.6	13.33, 20	0.08, 0.19, 0.29 (20 μm), and 0.285 (13.33 μm)	300,000
3D	1.8	13.33, 20	0.08, 0.19, 0.29 (20 μm), and 0.285 (13.33 μm)	170,000
3D	3.0	13.33, 20	0.08, 0.19, 0.29 (20 μm), and 0.285 (13.33 μm)	100,000
3D	12.0	13.33, 20	0.08, 0.19, 0.29 (20 μm), and 0.285 (13.33 μm)	25,000
3D	24.0	13.33, 20	0.08, 0.19, 0.29 (20 μm), and 0.285 (13.33 μm)	12,500
3D	48.0	13.33	0.285	6,250

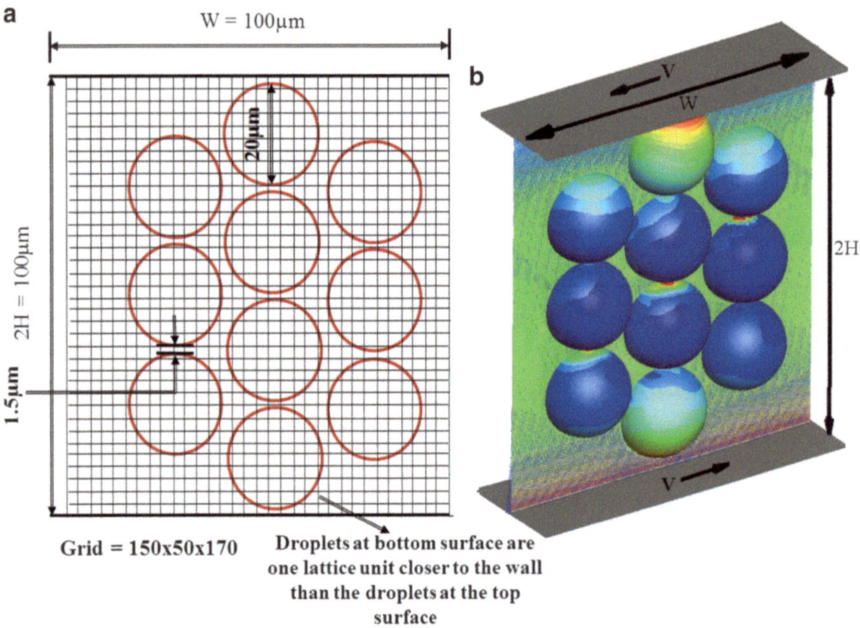

Fig. 5.6 (**a**) Schematic of the domain used for the simulation of emulsions under shear flow. It should be noted that the arrangement of droplets is in such a way that the droplets at bottom surface are closer to the wall by one lattice unit than the droplets at the top surface. (**b**) Phase field contours for two droplets in a simple shear flow with an indication of the used boundary conditions. The parameters used for the derivation of the effective viscosity are shown in the figure. The *top* and *bottom* walls are sheared in opposite directions with a velocity **V**. ***W*** is the length of the domain and **H** is the half width of the domain

in two opposite directions on top and bottom walls respectively as shown in Fig. 5.6. All cases were simulated till the effective viscosity is stabilized with time. This occurs when the deformation of droplets reaches a steady-state condition. The capillary number of droplets chosen for the present simulations was less than the critical value and thus no breakup of droplets is observed during the course of simulations. The maximum number of time steps simulated was 5,000,000 for 2D simulation at shear rate of 0.06 s^{-1}. Due to the computational cost involved, the present 3D simulations were only considered from shear rate of 0.6 s^{-1}, which needed 300,000 time steps to reach the steady-state condition.

5.3.2 *Influence of Flow Shear Rate on Aggregates of Suspended Droplets*

Numerous works in literature deal with the rheology and breakup of hard sphere colloidal aggregates under the influence of shear flow. It was observed that the higher shear rates are effective in breaking the colloidal aggregates (Tolpekin et al. 2004;

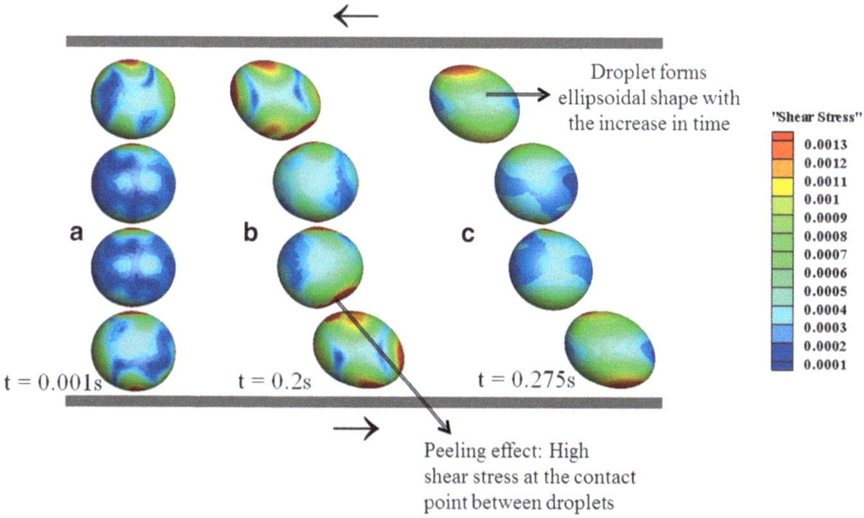

Fig. 5.7 Shear of four droplets at different instance of time is shown when the applied shear rate is 3 s^{-1}. The droplets are seen to align in the direction of flow with increase of time. Droplets close to wall deform and form an ellipsoidal shape which is typically observed for droplets under the influence of shear flow. The contours of droplet show the shear stress values, which are found to be maximum at the point of contact between two droplets indicating aggregation and peel of droplets under the influence of shear force

Derksen and Eskin 2010). However, the aggregates formed in emulsions have not been discussed widely. These aggregates are more complex due to their ability to deform under the flow conditions, thus causing the enhanced or suppressed deformation of droplets. In this section, the influence of flow shear rate on the behavior and breakup of aggregates of suspended droplet was studied. Understanding the behavior of aggregates will give us a better understanding to the rheology of emulsions.

To study the influence of shear flow on aggregates, four droplets were suspended in a single column formation and applied shear force on upper and the lower walls as shown in Fig. 5.7. Shear flow with shear rate ($\dot{\gamma}$) of 3 s^{-1} is used here. Due to the influence of shear force droplets close to the walls are found to peel off and align in the direction of flow. A maximum shear stress is found between the droplets at the area of contact, indicating that the peel off occurs locally at this location which initiates further breakup of aggregate droplets. This is evident in Fig. 5.7b, where the shear stress contours are shown over the droplet surface.

As the droplets move away from one another, the shear stress values at the area of contact are found to decrease and reach intermediate values (Fig. 5.7c) suggesting the disintegration of droplets. It can be observed that the droplets close to wall now slowly align in the direction of flow and form an ellipsoidal shape. This ellipsoidal shape is commonly observed with the droplets under the influence of shear flow (Farhat et al. 2011).

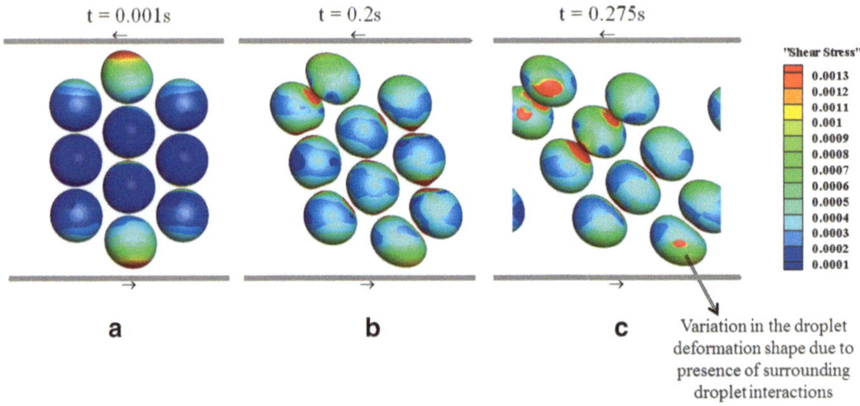

Fig. 5.8 Shear of ten droplets at different instance of time is shown when the applied shear rate is 3 s^{-1}

After simulating four droplets and observing that the droplet aggregation impedes the deformation of droplet, the aim was to see the effect of this aggregation for high volume fraction emulsions. Thus, ten droplets were arranged in three close columns with three, four, and three droplets respectively in each of the column, as shown in Fig. 5.8a. At the shear rate of $\dot{\gamma} = 3s^{-1}$, the droplets are observed to be slightly affected by the presence of adjacent droplets (Fig. 5.8b, c) and their deformation is slightly different from the deformation observed in a single column simulation. The deformation of droplets is observed to be significantly higher at the higher shear rate $\dot{\gamma} = 12s^{-1}$, and at the instance of disintegration, the droplets are observed to have formed into a shape distinctly different from the ellipsoid shape. The droplets are also observed to align in the direction of flow (Fig. 5.9).

To study the influence of aggregation quantitatively, change in viscosity of emulsions with time is plotted at different volume fractions and shear rates (Fig. 5.10). Viscosity of emulsions is observed to decrease with the increase in time. The reason for this reduction in the value of viscosity can be attributed to the de-aggregation of emulsion droplets. The evolution of viscosity with time can be differentiated into two phases. In the first phase, the particles de-aggregate under the influence of shear force and thus the viscosity decreases gradually. This is also the phase where droplets start deforming under the influence of shear flow. Due to an increase in deformation and the alignment of droplets along the flow field, the droplets become less resistive to the flow field. In the second phase, de-aggregation and deformation of the droplets reach a steady state; and this leads the viscosity of emulsions to reach the steady state condition. Viscosity of emulsions is found to be almost constant in this stage. This kind of evolution was previously observed by Chauveteau et al. (2000) and Omari et al. (2003) during the formation of microgels under the influence of shear flow.

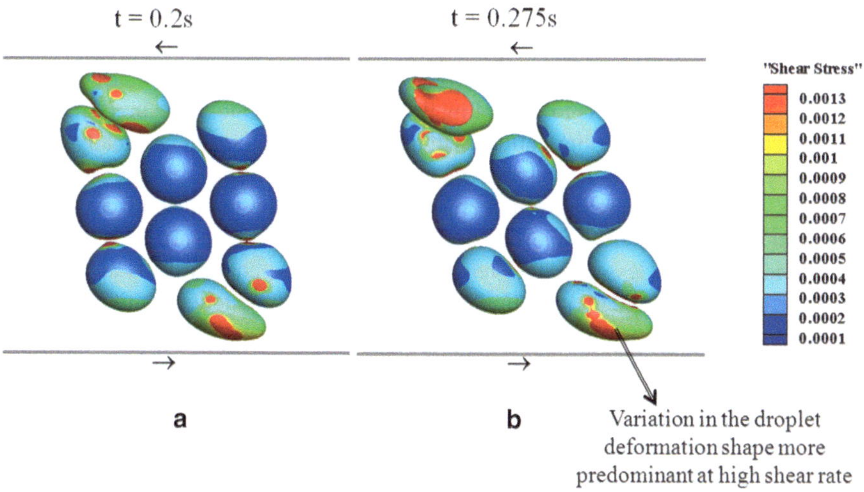

Fig. 5.9 Shear of ten droplets at different instance of time is shown when the applied shear rate is 12 s^{-1}

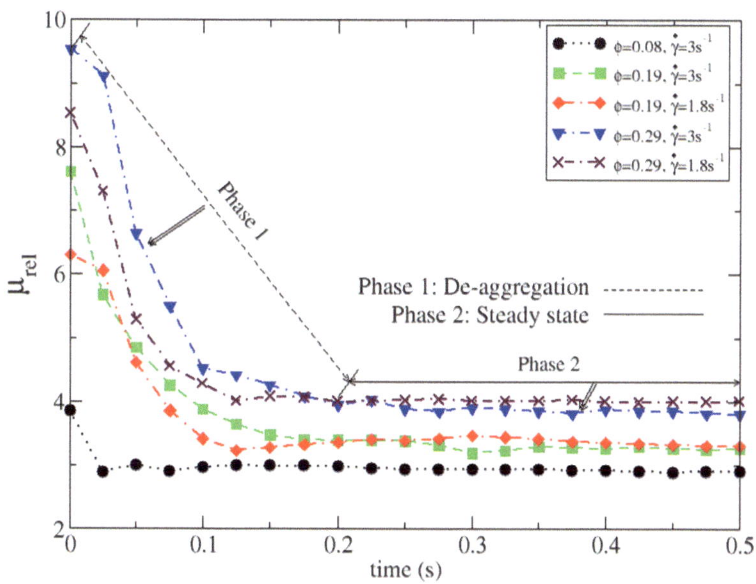

Fig. 5.10 Time evolution of relative viscosity of emulsions is shown here. Evolution takes place in phases of de-aggregation and steady state. De-aggregation phase increases with an increase in volume fraction. Time required for shear rates of 1.8 and 3 s^{-1} at volume fraction of 0.19 to reach steady state is 0.1 s and 0.175 s respectively. Time required for shear rates of 1.8 and 3 s^{-1} at volume fraction of 0.29 to reach steady state is 0.125 s and 0.12 s respectively. It is apparent that the lower shear rate reaches steady-state condition quicker than the higher shear rate

Figure 5.10 shows that the viscosity of emulsions is higher for low shear rates in the steady-state zone, which is expected from rheological curve of emulsions (Pal 2000, 2001). However, an interesting phenomenon observed from the plot is that the time spent in phase 1 is lower at low flow shear rates. Due to the high shear rate of flow, higher shear rate simulations will de-aggregate and reach the steady state at a faster rate. But at both the volume fractions of $\phi = 0.19$ and 0.29, the shear rates of 1.8 s^{-1} reached steady state quicker than the 3 s^{-1}. The conclusion from this plot is that at low shear rates, droplets do not de-aggregate as effectively as at high shear rates and thus need lesser time to reach the steady-state condition. Thus the close packing of droplets and their motion as a bulk fluid lead to higher relative viscosity during the steady-state phase.

From the study, one can also conclude that the droplet aggregations affect their deformation and influence the effective viscosity of the fluid. Another observation is that at higher volume fractions, the aggregated droplets do not disintegrate easily and thus move in the flow as a bulk formation which could be one of the reasons why high viscosity is found at higher volume fractions.

5.3.3 *Shear Rate Dependent Viscosity*

The relative viscosity ($\mu_{rel} = \mu_{eff}/\mu_{fluid}$), the ratio of the effective viscosity to fluid viscosity, is shown for different shear rates in Fig. 5.11. The plot compares the effective viscosity values for 2D and 3D simulations. The plot indicates that the effective viscosity decreases with an increase of shear rate. Similar phenomenon is observed in all the volume fractions. The results are consistent with the general trend of fluids with non-Newtonian characteristics. A closer look at the plot reveals that the 2D simulations show a shear thinning behavior with shear thinning index of n ~ 0.74, where $\mu_{rel} \propto \gamma^{\bullet}\ (n-1)$. Here $\dot{\gamma}$ is shear rate of the flow. The shear thinning behavior for 3D simulations is observed to have shear index ranging from n = 0.88 to 0.94. Similar shear thinning index of n = 0.9 for simple emulsions was also observed by Peixinho et al. (2010). The reason for larger shear thinning in 2D simulations is due to larger droplet deformation predicted by 2D simulations.

Cross model (Escudier et al. 2001) (Eq. 5.10) estimation of 3D simulations at all three volume fractions is plotted and shown in Fig. 5.12. The model was plotted from $\dot{\gamma} = 0.01 - 100s^{-1}$ to observe the extended behavior of the emulsion rheology. It can be observed that the Cross model shows a good prediction to simulation results with shear thinning index of $q = 0.8$. The model indicates that the difference between μ_{0rel} and $\mu_{\infty rel}$ is small, thus suggesting a weak shear thinning phenomenon. This behavior was consistent for all three volume fractions. The parametric constants in these correlations are dependent on many different variables such as surfactant composition, surfactant concentration, and volume fraction of emulsions. This makes it difficult to determine the values of parametric constants which limit

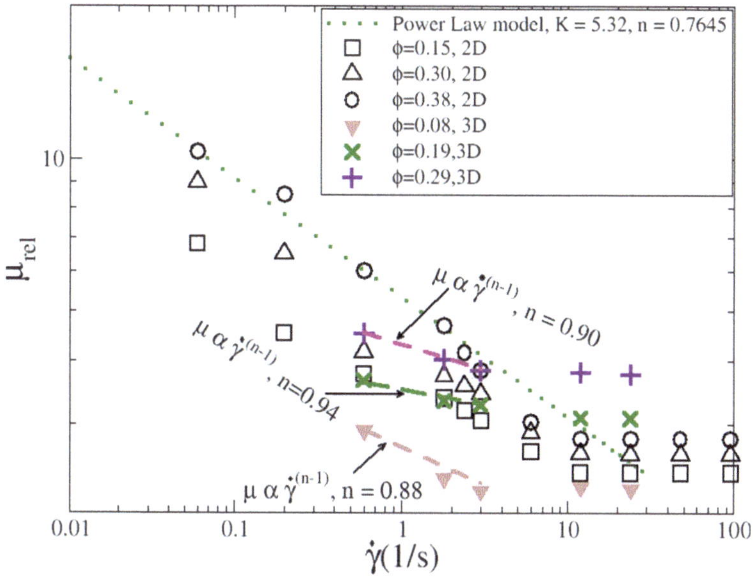

Fig. 5.11 Relative viscosity is plotted for different applied shear rates at different volume fractions for both 2D and 3D simulations. The relative viscosity is observed to decrease with an increase of the flow shear rate. The shear thinning index (n) is observed to be approximately equal to 0.9 for 3D simulations whereas for 2D simulations n is found to be equal to 0.76. The reason for larger slope of shear thinning for 2D simulations is due to the large deformation prediction of droplets. Power law and Cross model Escudeir et al. (2001) are also plotted here. Both the models involve parametric constants which are difficult to determine for an individual flow characteristics

the use of empirical models in predicting the shear rate-dependent relative viscosity.

$$\mu_{rel} = \mu_{\infty rel} + \frac{\mu_{0rel} - \mu_{\infty rel}}{1 + (C\dot{\gamma})^q} \tag{5.10}$$

At smaller shear rates, the droplets move as bulk aggregates as has been discussed earlier and thus increase the effective viscosity. At higher shear rates the droplets quickly disintegrate and align in the flow direction, thus reducing the resistance to the fluid flow. Therefore the effective viscosity is reduced at high shear rates. As the droplet deformation becomes steadier with an increase in shear rate, the emulsions tend to show Newtonian characteristics.

In order to study the effect of initial droplet position on the viscosity of emulsions, $\phi = 0.08$ was simulated while placing the droplets diagonally inside the computational domain such that the droplets are far off from each other initially. The initial arrangement of such simulation is shown in the inset of Fig. 5.13 with a header "Diagonal." Three simulations were performed for this initial position with shear rates of $\dot{\gamma} = 0.6, 3.0, 12.0s^{-1}$. The maximum difference between the viscosity

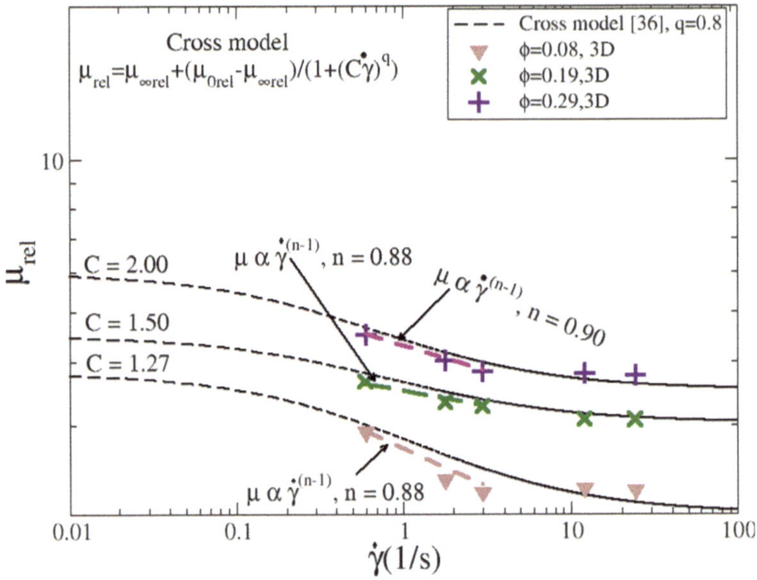

Fig. 5.12 Relative viscosity is plotted for different applied shear rates at different volume fractions for 3D simulations and is compared with the Cross model. Cross model shows a good prediction with the simulated data of the shear thinning index value of q = 0.8. The range of difference between high and low viscosity values of same volume fraction is low indicating a low shear thinning behavior

of emulsion when droplets are arranged in diagonal and vertical (initial position is shown in Fig. 5.13 with a header "Vertical") is found to be 2.25 %. These simulations indicate that the effect of initial droplet position on the viscosity of emulsions calculated at the end of the simulation is negligible.

Figure 5.13 also shows the rheological curves for droplets of 13.33 and 20 μm with their respective being $\phi = 0.285$ and $\phi \approx 0.29$. The plot indicates that the smaller size of droplets shows higher viscosity at low shear rates. Also the shear thinning behavior for droplets with size 13.33 μm was observed to be greater (n = 0.79). This variation in viscosity values with the change in droplet size was experimentally observed by Pal (2000) for petroleum oil/water emulsion system. The reason for this change in viscosity values should be due to the fact that increase in number of droplets increases the aggregational and repulsive forces within the emulsion system.

To quantify the deformability of droplets, the average deformation index of all droplets at $\phi = 0.19$ was plotted for the range of shear rates where shear thinning is observed ($\dot{\gamma} = 0.6 - 3.0s^{-1}$). The deformation index of droplet is the measure of deformation of droplet and is given as

$$DI = \frac{L - B}{L + B} \tag{5.11}$$

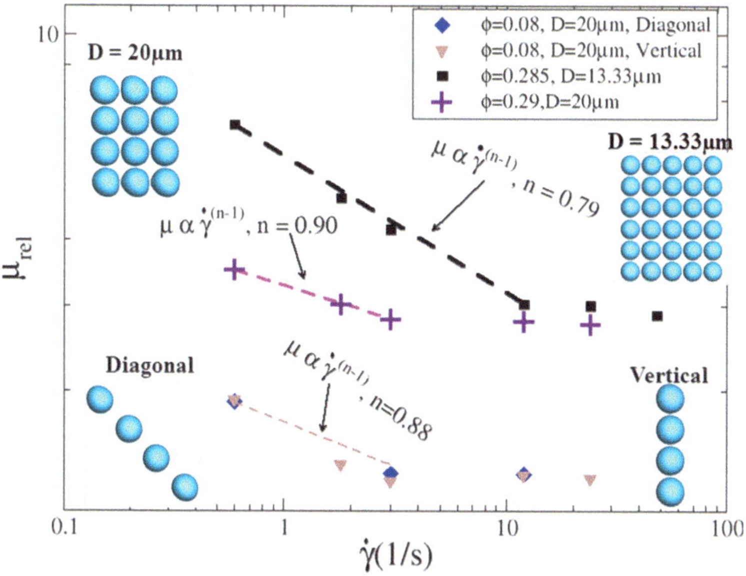

Fig. 5.13 Relative viscosity is plotted for different applied shear rates at volume fractions of 0.08 and 0.29 for 3D simulations. Simulations performed at $\phi = 0.08$ are used to study the effect of position of droplet on the final viscosity calculation. *Insets* with headers "Diagonal" and "Vertical" are two initial positions of droplets used for the study. The simulations performed with $\phi \approx 0.29$ are used to study the effect of droplet size on the rheological characteristics of emulsions. The *insets* shown with headers "D = 13.33 μm" and "D = 20 μm" indicate the initial setup of droplets used in this simulation

where DI is called the deformation index, L is the maximum length of the droplet, and B is the maximum width of the droplet. L and B values used for this study have been directly measured from the simulations. The average deformation index (DI_{ave}) value is observed to increase with increase of shear rate (Fig. 5.14), suggesting that the deformation of droplet increases with an increase in shear rate. It can be visually verified from Fig. 5.15, where the droplets isocontours are shown at the end of the simulation for $\phi = 0.19$ and shear rates ranging from $\dot{\gamma} = 0.6 - 3.0s^{-1}$.

It should be noted that the shear rate for DI_{ave} is plotted in reverse direction (Fig. 5.14). This has been done to be able to compare the slope of shear thinning of μ_{rel} to the slope of increase in DI_{ave} in the same region. The average slope of shear thinning curve is found to be almost parallel to the slope of increase in DI_{ave}. It suggests that the deformation of droplets and alignment of droplets (which is measured by the angle of major axis L with the horizontal direction) are chiefly responsible for the shear thinning behavior (see Fig. 5.15 and Movies C5.3.3—Rheo_10drops_0.6_per_sec, C5.3.3—Rheo_10drops_1.8_per_sec and C5.3.3—Rheo_10drops_3.0_per_sec). The small discrepancies in the slopes of shear

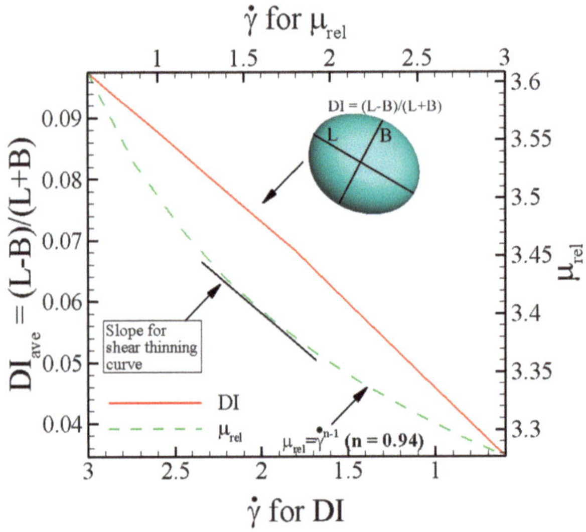

Fig. 5.14 The plot represents the deformation index and relative viscosity of emulsions with the change of shear rate from range of 0.6 to 3 s^{-1}. The parameters of deformation index and relative viscosity are shown here for the volume fraction of $\phi = 0.19$. Deformation index is observed to increase with an increase of shear rate indicating greater deformation at larger shear rates. Due to large deformation, the relative viscosity of emulsions is observed to decrease with shear rate. The average slope of shear thinning curve is almost parallel to the slope of deformation index suggesting a clear correlation between them

thinning curve of μ_{rel} and that of DI_{ave} could be due to the aggregation characteristics of droplets which also play a role in the non-Newtonian characteristics of emulsions.

Rheological behavior of emulsions can also be studied by plotting shear stress against shear rate. The relative shear stress (σ) was plotted against shear rate on a log–log plot to study this behavior (Fig. 5.16). Relative shear stress is calculated by multiplying the relative viscosity of emulsions with the corresponding shear rate. Figure 5.16 shows that the relative shear stress increases with an increase in the volume fraction. The relative shear stress is also observed to increase gradually with an increase in the shear rate. Yield stress behavior is not observed in Fig. 5.15 as the shear rates considered in this study occur after the yield shear point. Considering shear stress proportional to power of shear rate ($\sigma \propto \gamma^{\bullet}$ m), the values of m were found to be equal to 1.045, 1.072, and 1.052 for volume fractions of 0.08, 0.19, and 0.29, respectively. The values of power index suggest an almost linear relationship between shear stress and the shear rate. However, a slight convex nature of shear stress to shear rate plot implies that the emulsion behaves like a pseudo-plastic fluid. It should be noted that the index values of m are different from n index values taken from the viscosity curves. This is because the m index was calculated for shear rates ranging from $\dot{\gamma} = 0.6 - 24.0s^{-1}$. The n index values were only computed for shear rates where shear thinning behavior was apparent. Weak

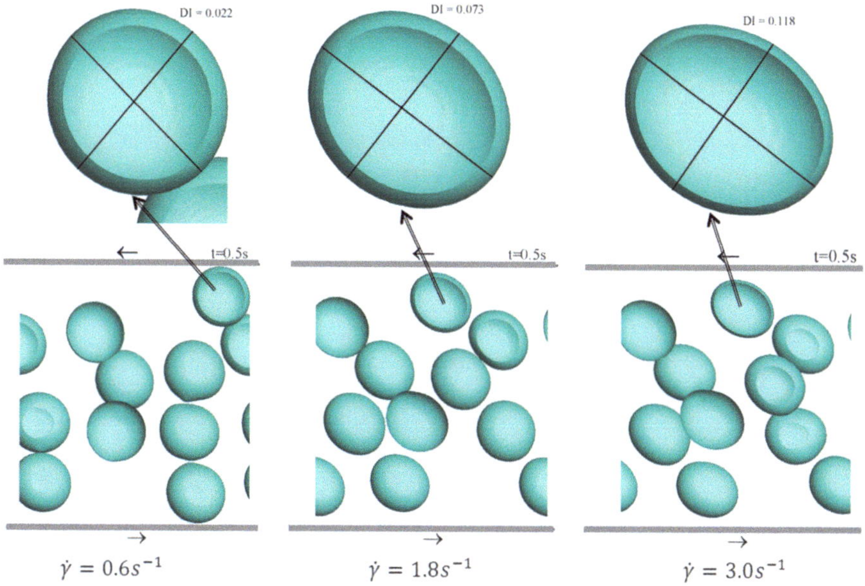

Fig. 5.15 Droplets at three different shear rates are shown at the end of the simulation. Clearly the deformation is larger for shear rate of 3 s^{-1} when compared to 1.8 s^{-1}, which in turn has larger deformation than at shear rate of 0.6 s^{-1}. Deformation index values of a single droplet are calculated and shown in the figure. DI of the droplet at shear rate of 3 s^{-1} = 0.118; at shear rate of 1.8 s^{-1} = 0.073; and at shear rate of 0.6 s^{-1} = 0.022

shear thinning behavior observed from viscosity curve and shear stress curve is due to the low concentration of emulsions, where droplets quickly deform and achieve a stable shape under the influence of shear flow.

5.3.4 *Volume Fraction Effects*

An increase in the relative viscosity of emulsions was observed with an increase in the volume fraction (Fig. 5.17). The result is qualitatively consistent with the theoretical and experimental data of previous studies (Pal 1996, 2000; Mason et al. 1996; Ladd 1990). It was found that the difference in the relative viscosity between different volume fractions is observed to be greater at low shear rate and it gradually reduces at higher shear rates (Fig. 5.17). It was previously observed (Fig. 5.7) that at low shear rate the droplet aggregates move as bulk fluid before disintegrating into individual droplets. Thus, the larger volume fractions create a larger portion of bulk formation which should directly influence the relative viscosity at different volume fractions. Whereas at very high shear rates, due to a quick disintegration of droplets and strong alignment in the flow direction, different volume fractions show almost similar resistance to the flow field. It is therefore

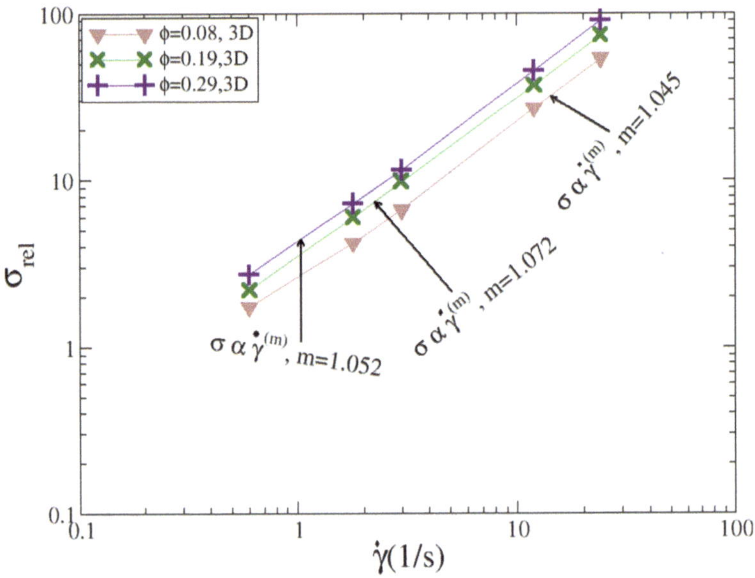

Fig. 5.16 Plot for relative shear stress of emulsions for different shear rates is shown here. The plot shows an increase in the shear stress with increase of shear rate and volume fraction. The shear rates considered for the simulation start after the yield shear stress point. Increase in shear stress values is almost linear indicating that there is a weak shear thinning behavior for the emulsions considered in the simulation

observed that the difference between the relative viscosities at different volume fractions is smaller at higher shear rates.

Figure 5.17 shows the relative viscosity of emulsions for both 2D and 3D simulations. With an increase in the volume fraction the slope of increase in relative viscosity of 2D simulations is small at high shear rates. This is due to the fact that at high shear rates the deformation of the droplets in 2D simulations is large which causes the low increase in the relative viscosity. The plot also shows the experimental data (Chandran et al. 2007) of relative viscosity at different volume fractions. Though the shear rate and viscosity ratio of droplets to emulsions (λ) are slightly different, the range of 3D simulation data investigated in this study show the same trend for the slope of increase of relative viscosity as observed from the experimental data. It is right time to mention that the empirical correlations (Pal 2000; Hsueh and Wei 2009; Bedeaux 1983; Hashim and Shtrikman 1963; Phan-Thien and Pham 1997) used to estimate the relative viscosity of emulsions at different volume fractions are independent of shear rate. From the simulations, it was found that the relative viscosity significantly varies with the change in shear rate. Thus, neglecting the effect of shear rate in the correlations can be considered as one of the major drawbacks of empirical models used to estimate the relative viscosity of emulsions at different volume fractions.

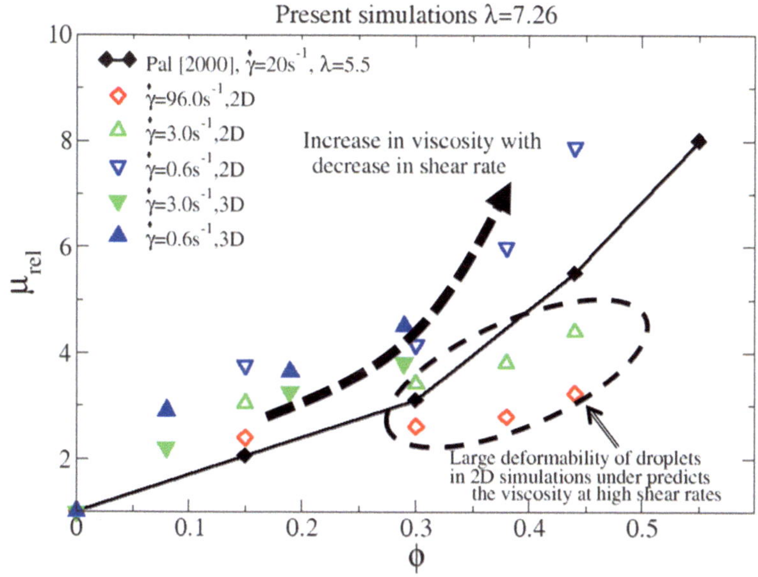

Fig. 5.17 Relative viscosity as a function of volume fraction is shown for different shear rates. The plots suggest an increase in the relative viscosity with an increase of volume fraction. 2D simulations underpredict the relative viscosity of emulsions at high shear rates due to large deformation of droplets. 3D simulations of relative viscosity show a slope similar to that found in the experiments

Chapter 6
Nonuniform Interfacial Tension LBM for RBC Modeling

Abstract This chapter aims at analyzing the red blood cell (RBC) deformation and velocity while streaming through venules and through capillaries whose diameters are smaller than the RBC size. The characteristics of the RBC shape change and velocity can potentially help in diagnosing diseases. In this work the RBC is considered as a surfactant covered droplet. This is justified by the fact that the cell membrane liquefies under pressure in the capillaries, and this allows the marginalization of its mechanical properties. The RBC membrane is in fact a macrocolloid containing lipid surfactant. When liquefied, it can be considered as a droplet of immiscible hemoglobin covered with lipid surfactant in a plasma surrounding. The local gradient in the surface tension due to nonuniform local interface surfactant distribution is neglected here, and a nonuniform zonal averaged value of surface tension representative of the surfactant bulk zonal concentration is rather implemented. The interplay between the surface tension geometry and the hydrodynamic conditions determines the droplet shape by affecting a change in its Weber number, and influences its velocity. The Gunstensen lattice Boltzmann model for immiscible fluids is used since it provides independent adjustment of the local surface tension, and allows the use of fluids with viscosity contrast. The proposed concept was used to investigate the dynamic shape change of the RBC while flowing through the microvasculature, and to explore the Fahraeus, and the Fahraeus–Lindqvist effects.

This chapter aims at analyzing the red blood cell (RBC) deformation and velocity while streaming through venules and through capillaries whose diameters are smaller than the RBC size. The characteristics of the RBC shape change and velocity can potentially help in diagnosing diseases. In this work the RBC is considered as a surfactant covered droplet. This is justified by the fact that the

---

Electronic supplementary material Supplementary material is available in the online version of this chapter at http://dx.doi.org/10.1007/978-1-4899-7402-0_6. Videos can also be accessed at http://www.springerimages.com/videos/978-1-4899-7401-3

H. Farhat et al., *Accelerated Lattice Boltzmann Model for Colloidal Suspensions: Rheology and Interface Morphology*, DOI 10.1007/978-1-4899-7402-0_6,

cell membrane liquefies under pressure in the capillaries, and this allows the marginalization of its mechanical properties. The RBC membrane is in fact a macrocolloid containing lipid surfactant. When liquefied, it can be considered as a droplet of immiscible hemoglobin covered with lipid surfactant in a plasma surrounding. The local gradient in the surface tension due to nonuniform local interface surfactant distribution is neglected here, and a nonuniform zonal averaged value of surface tension representative of the surfactant bulk zonal concentration is rather implemented. The interplay between the surface tension geometry and the hydrodynamic conditions determines the droplet shape by affecting a change in its Weber number, and influences its velocity. The Gunstensen lattice Boltzmann model for immiscible fluids is used here since it provides independent adjustment of the local surface tension, and allows the use of fluids with viscosity contrast. The presented concept was used to investigate the dynamic shape change of the RBC while flowing through the microvasculature, and to explore the Fahraeus, and the Fahraeus–Lindqvist effects.

6.1 The Heuristic Approach for Surfactant-Covered Droplets

The motivation for the use of zonal averaged nonuniform surface tension method stems from the fact that the lipid membrane liquefies at pressure greater than 29 dyne/cm as explained by Keller et al. (1998), a condition sought to prevail in the microvasculature. Thus, the liquefied lipid bilayer is expected to allow its surfactant molecules to shift in the direction opposite to the flow direction mainly due to convection which is dependent on the tangential surface velocities. Since the number of lipid molecules is constant on the periphery of the RBC and since these molecules are finite by size, the shift comes to a halt when regions of higher surfactant concentration reach their saturation point. Another important factor which prevents any further buildup of the concentration is the Marangoni stress which is expressed as follows:

$$-\nabla_s \sigma_s = -\frac{\partial \sigma_s}{\partial \Gamma} \cdot \nabla_s \Gamma \tag{6.1}$$

where ∇_s is a surface gradient, σ_s is the local surface tension, and Γ is the local surfactant concentration. It is convenient to relate the surface tension to the surfactant concentration by the Langmuir surface equation of state expressed as follows:

$$\sigma_s = \sigma_0 + RT\Gamma_\infty \ln(1 - \Gamma/\Gamma_\infty) \tag{6.2}$$

where σ_0 is the surface tension of a clean surface, Γ_∞ is the saturated surfactant surface concentration, R is the gas constant, and T is the temperature in Kelvin.

Equations (6.1) and (6.2) yield the following surface tension derivative with respect to the surfactant concentration (Hu and Lips 2003):

$$\frac{\partial \sigma_s}{\partial \Gamma} = -\frac{RT}{(1-\Gamma/\Gamma_\infty)} \tag{6.3}$$

Equation (6.3) indicates that an increase in the surfactant surface concentration leads to an increase in the Marangoni stress, which in turn slows the tangential surface velocity and hampers any further convection of surfactant towards the regions of higher concentration. This process prevents the presence of regions with extremely low surface tension and limits the area generation besides the limitations imposed by the contraction of the regions with lower surfactant due to higher surface tension. Equation (6.2) can be written in the following dimensionless form:

$$\frac{\sigma_s}{\sigma_0} = 1 + E_0 \ln(1-\Gamma/\Gamma_\infty) \tag{6.4}$$

where $E_0 = RT\Gamma_\infty/\sigma_0$ is the surfactant elasticity which determines the strength of the surfactant effect on the interfacial tension.

In this model an estimated value of $E_0 = 0.2$ was used. A reasonable range for the bulk surfactant concentration ratio of $0.3 \leq \Gamma/\Gamma_\infty \leq 0.7$ was considered in Eq. (6.4) based on a study by Hu and Lips (2003) for some polymer blends. An initial surface tension parameter α_0 from Eq. (2.16) was set such that the droplet deformation index $DI \approx L/D$ was equal to an experimental RBC index, where L, D were the length and the width of the RBC (Hong Jeong et al. 2006). An initial surfactant concentration ratio $\Gamma_i/\Gamma_\infty = 0.7$ was used with the assumption that the lipid bilayer cannot be fully saturated throughout the RBC membrane as stated by Braasch (1971) regarding the radio-autography results. For simplicity the interface of the droplet was then divided into two adjustable regions: frontal region with lower surfactant concentration and backside region with higher concentration as shown in Fig. 6.1, such that the initial total surfactant mass was conserved on the droplet interface:

$$\begin{aligned} &\frac{\Gamma_b}{\Gamma_\infty} N_b + \frac{\Gamma_f}{\Gamma_\infty} N_f = \frac{\Gamma_i}{\Gamma_\infty} N_T \\ &\frac{\Gamma_f}{\Gamma_\infty} \leq \frac{\Gamma_b}{\Gamma_\infty} < 1 \end{aligned} \tag{6.5}$$

where N_b, N_f, N_T are the number of interfacial nodes in the back, front, and total regions respectively. Γ_f, Γ_b, Γ_i are the frontal, backside, and initial surfactant concentration. The zonal surface tension parameters of the model had to obey the following equation:

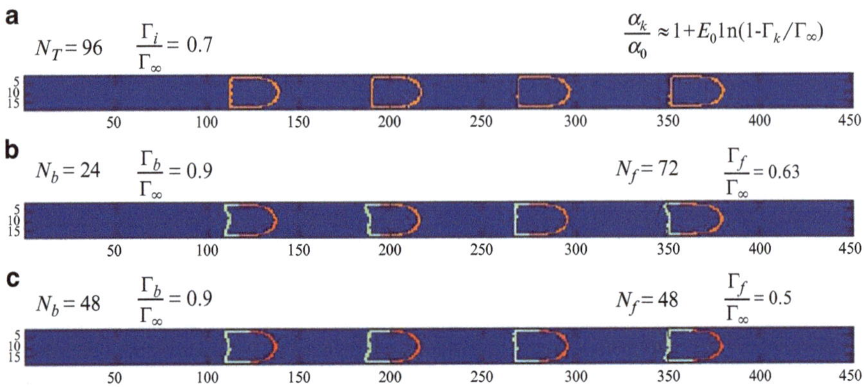

Fig. 6.1 Illustrates the adjustable zonal division of the interface with regions of higher (frontal) and lower (backside) interfacial tension. The *darker* contours show greater surface tension parameter α from Eq. (2.16). (**a**) Shows homogeneous α value, (**b**) quarter of the interface had lower α value; (**c**) the interface was shared equally between high and low values of α

$$\frac{\alpha_k}{\alpha_0} \approx \frac{\sigma_k}{\sigma_0} = 1 + E_0 \ln(1 - \Gamma_k/\Gamma_\infty) \tag{6.6}$$

where $k = f, b$ refer to the frontal and backside regions respectively.

A viscosity ratio $\mu_h/\mu_p = 7$ was used throughout this model which is an approximation of the ratio of hemoglobin to plasma. Preferential wetting was given to the suspending fluid which was treated as hydrophilic, contrary to the droplets which were hydrophobic.

6.2 Simulation Results and Discussion

6.2.1 *Surface Tension Effects on a Single File Flow RBCs Shape and Velocity*

To calibrate the model data were extracted from the experimental work of Hong Jeong et al. (2006) which indicated RBC deformation index $DI = 1.55$ and velocity $V = 1.8mm/s$ inside the capillary of a rat mesentery. Assuming that blood density at normal temperature is $\rho = 1.05g/cm^3$ as was stated by Nakano et al. (2005) and that the velocity of the RBCs is representative of the average blood velocity in the capillary, the estimated blood viscosity is $\mu = 2.35cP$ and the average Reynolds number is $\mathrm{Re} \approx 0.0055$ for a random capillary size.

A domain of 19×450 lattice sites was selected for the simulation of four droplets where all variables were measured in lattice units. The relaxation times were chosen as $\tau_p = 0.602$; $\tau_h = 1.219$; $\tau_{eff} = 0.908$ for the suspending, the suspended fluid, and the interface as per Eq. (2.15), respectively. This produced the right viscosity ratio between the two fluids and an apparent kinematic viscosity $\nu_{app} \approx 0.08$ which was calculated based on the density fraction of the red and blue fluids. A density $\rho = 1.05$ was used for both fluids in the model. The average velocity of the bulk fluid was calculated using the known experimental value of the Reynolds number $\mathrm{Re} = V_{av}2h/\nu_{app} \approx 0.0055$, where h was half the channel width. This led to an average velocity $V_{av} \approx 2.5 \times 10^{-5}$ from which the source term was derived as follows (Sukop and Thorne 2006):

$$F = \frac{2\rho\nu_{app}V_{av}}{h^2} = 5.8 \times 10^{-8} \tag{6.7}$$

The source term in Eq. (2.20) produced the right undisturbed average velocity of the bulk fluid, but since the droplets caused little resistance to the flow, an increase of the source term was deemed necessary, thus a value of $F = 7.8 \times 10^{-8}$ was found by trial to produce the needed average velocity. To maintain a deformation index $DI \approx 1.55$ and the calculated RBC average velocity, the following value for the surface parameter was found by trial $\alpha_0 \approx 1.5 \times 10^{-7}$ as shown in Fig. 6.2a. Periodic boundary conditions were applied at the inlet and outlet boundaries, and bounce-back conditions at the upper and lower walls. After determining the optimum surface tension parameter α_0, the zonal averaged nonuniform method with values calculated by Eqs. (6.5) and (6.6) was used in two cases where a lower surface tension parameter value covered about quarter of the interface in one case and half of the interface in the other. The influence of the nonuniform surface tension was more noticeable on the droplets shape through the appearance of tails due to the lower surface tension at the backside and the influence of the shear stresses which were higher near the walls. A velocity drop was also witnessed in case (B) and case (C). This could be due to a relatively higher surface tension parameter resulting from Eqs. (6.5) and (6.6) in the frontal region which led to a slightly lower droplet average velocity. In the subsequent cases the surfactant concentration distribution of case (B) will be used as a reference.

More simulations were performed using the same domain and flow conditions, but the surface tension parameter α_0 was varied in each run. The intention was to investigate the range of the surface tension parameters responsible for slowing down the droplet to about 20–0 % of its nominal velocity and see its effects of the RBC shape.

The mean velocity of the RBC was calculated using the following equation:

$$\boldsymbol{u}_{av} = \frac{1}{A}\iint \boldsymbol{u}_x dxdy \tag{6.8}$$

where A is the total area of the RBCs and the deformation index was computed by the following formula:

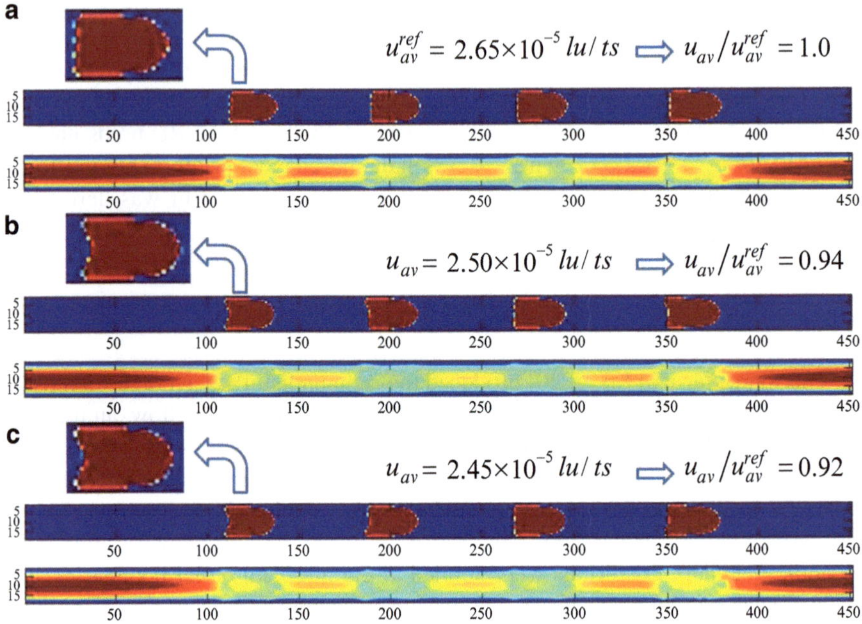

Fig. 6.2 Phase field and horizontal velocity contours for four droplets streaming in a narrow vessel with homogeneous surface tension parameter (**a**), and with zonal averaged nonisotropic surface tension parameter whose values were explained in Fig. 6.2 for *insets* (**b**) and (**c**)

$$DI = \frac{(a-b)}{(a+b)} \tag{6.9}$$

where a, b are the major and minor axis of the droplets respectively. Figure 6.3 shows the phase field and velocity contours for four runs where the surface tension parameter was varied $1.5 \times 10^{-7} \le \alpha_0 \le 1.05 \times 10^{-6}$ by an equal step.

The normalized average velocities of the four cases in which the reference case velocity (A) was used in the denominator and the DI were plotted in Fig. 6.4. A general trend of increasing velocity and DI is observed with decreasing surface tension parameter. It can be seen from Fig. 6.4 that the surface tension parameter had to be raised seven times with respect to the reference parameter in order to reduce the droplet velocity up to 20 % of its nominal value and the DI by 50 %.

Driessen et al. (1980) tested the effect of injecting hardened RBCs inside the rat mesentery. A decrease of the RBC velocity up to (71.1 ± 19.1)% was observed when RBCs treated with $0.5 mmol.L^{-1}$ diamide were placed in the rat capillaries during the exchange period and to (8.3 ± 15.1)% during the hypotension period. When $1.5 mmol.L^{-1}$ diamide was used the velocity dropped to (86 ± 24)% during the exchange period and to (2.1 ± 6.1)% during the hypotension period.

It is noticed from Fig. 6.4 that for $4.5 \times 10^{-7} \le \alpha_0 \le 7.5 \times 10^{-7}$ the normalized velocity curve had a gentle slope and the decrease in velocity was up to 65 %,

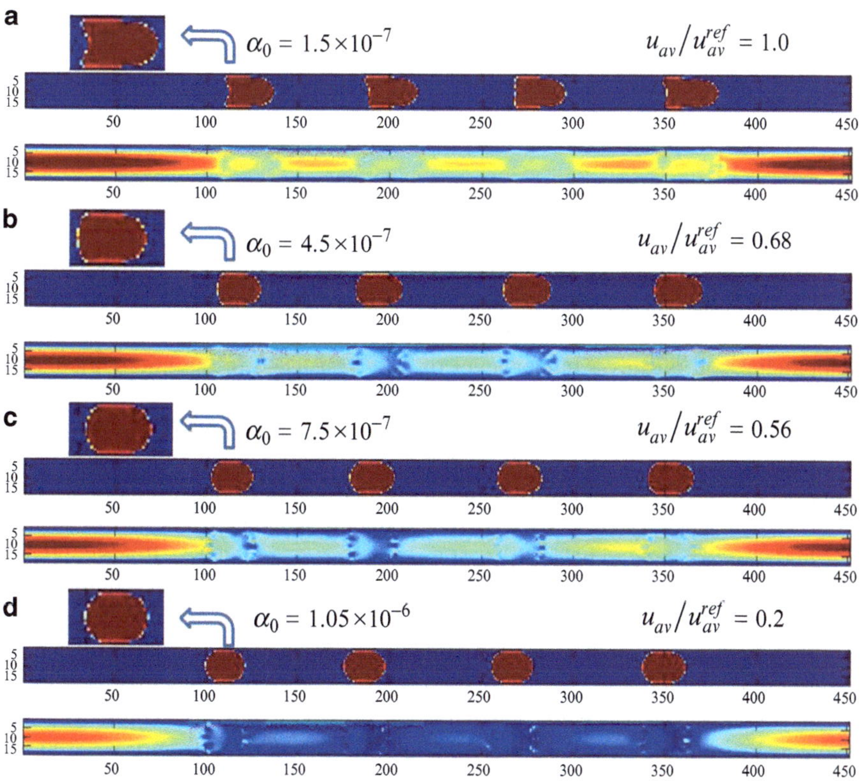

Fig. 6.3 Phase field contours and their respective horizontal velocity contours for surface tension parameters $1.5 \times 10^{-7} \leq \alpha_0 \leq 1.05 \times 10^{-6}$ by a step of $\delta_{\alpha_0} = 3.0 \times 10^{-7}$

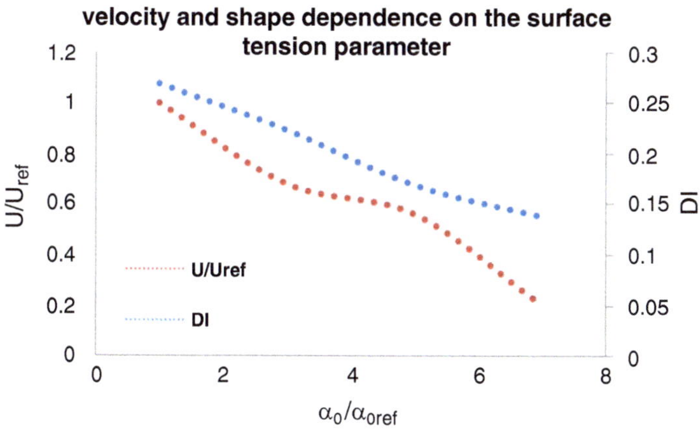

Fig. 6.4 Droplet velocity and DI dependence on the surface tension parameter

which agrees qualitatively with the experimental trend. This could pave the way for the potential of qualitatively correlating the model results with some experimental data such as the results of Driessen et al. (1980) if more details were available.

6.2.2 RBC Deformability

The role of the endothelial surface layer on the RBC deformability, flow resistivity, and lower tube hematocrit was studied by (Secomb et al. 1998, 2001). Special attention was given to the exclusion of the RBC from the capillary walls. The endothelial cell creates a higher level of flow resistance in the microvasculature compared to glass tubes used in in vitro experiments.

A domain consisting of 160×24 lattice units was used to model single RBC flowing in a capillary of 6 μm diameter. The results of this simulation were expressed in physical values to enable a comparison with the results of Secomb et al. (2001). The reference shape, before applying the nonuniform surface tension method, was assumed to be a sphere of 6 μm diameter as stated by Braasch (1971). The source term was varied in order to achieve droplet velocities ranging from $500\mu m/s$ to $3,700\mu m/s$ using Eq. (6.7). In the presented model velocities below $500\mu m/s$ were not used. This was to avoid lattice pinning, which originates from the Gunstensen method for the segregation step at very low velocities (D'Ortona et al. 1995). The surface tension parameters used in this simulation were the same as the reference case and were based on values derived from Eqs. (6.5) and (6.6). The selection ensured a good control over the shape of the RBC. The gap width between the RBC and the wall was measured in the same manner used by Secomb et al. (1998). The gap was considered as the average width along the whole length of the cell, where the angle between the tangent to the curvature of the cell and the vessel wall was less than 11° as shown in Fig. 6.5.

Figure 6.6 shows the droplet-wall gap width variation with respect to changing the velocity. The results obtained from the presented model were compared to the results of Secomb et al. (2001). The upper higher graph represents the model in which Secomb considered the role of the endothelial surface layer, and the lower graph corresponded to the model in which he disregarded that role. It is evident that the bounce-back condition in the present model did not fully recover the effects of the endothelial surface layer on the RBC exclusion from the wall, but it still produced reasonable results.

The droplet shape change due to the increase in its velocity is shown in Fig. 6.7. This was done by measuring the deformation index as per Eq. (6.9). The same graph shows also the droplet steady-state length and gap. A trend of increasing gap, length, and DI is associated with an increase in the droplet velocity.

In the presented model a relatively higher velocity was required to achieve a droplet elongation to about 8 μm as shown in Fig. 6.7. This was due to a smaller initial reference shape. The elongation of the droplet was compensated by a greater droplet-wall gap width.

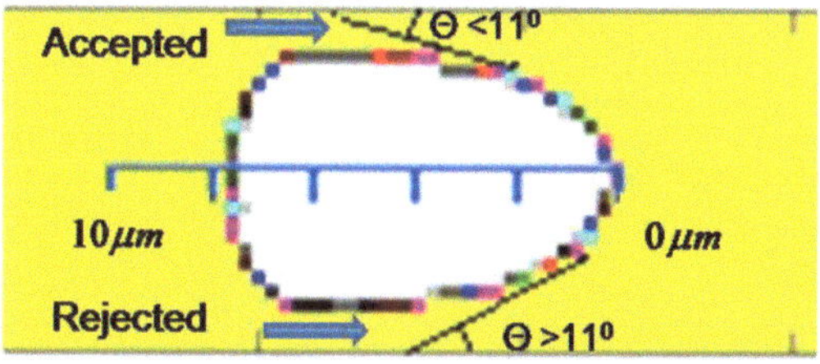

Fig. 6.5 Criteria for accepting or rejecting the gap measurement along the length of the RBC used for calculating the average width. For $\theta < 11°$ the record is taken; otherwise it is rejected

Fig. 6.6 Droplet-wall gap width comparison of the presented model with the data generated from Secomb et al. (2001) (Glycocalyx indicates that the endothelial cell layer was considered in Secomb's model and no glycocalyx indicates otherwise)

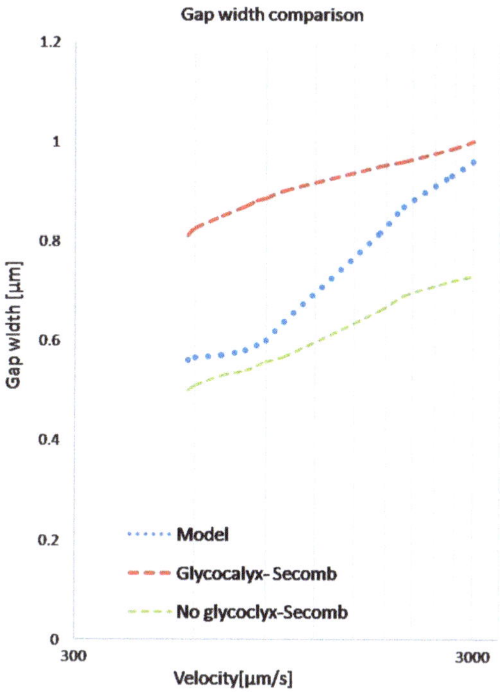

It is reasonable to assume that the deformation was caused mainly by the higher viscous stress near the walls since in the present model the droplet was not allowed to wet the walls. In a straight section of the microvasculature, the RBC tends to form a bullet shape when flowing at its normal velocity as shown in Fig. 6.8. This is due to the Poiseuille nature of the flow, which enables the forehead of the RBC to have higher momentum; meanwhile lower momentum prevails at the upper and lower sides because of the retardation caused by the viscous stresses. This results in an exclusion of the RBC from the walls and leads to a considerable reduction of the viscous stresses, and thus to the vanishing of its parachute since the surface tension effects overcame the shear stress effects.

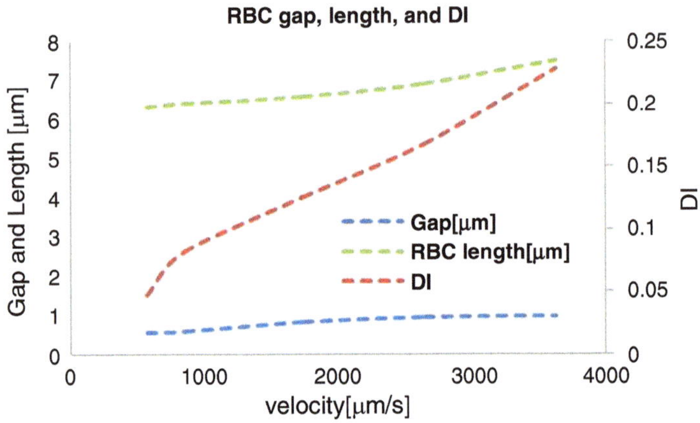

Fig. 6.7 Droplet-wall gap, DI, and length calculate with respect to the droplet velocity

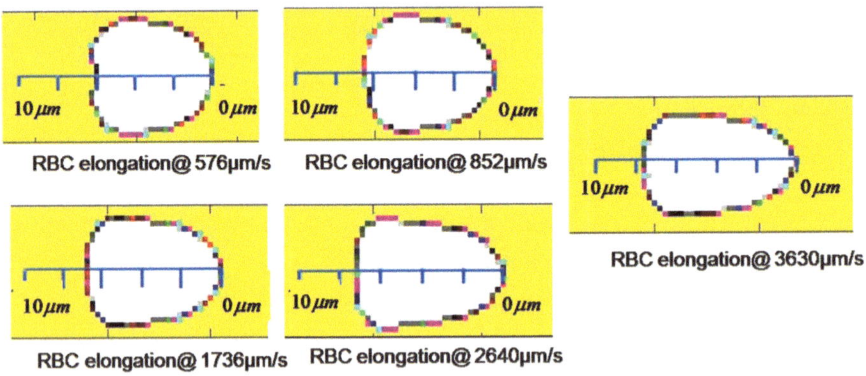

Fig. 6.8 Dependence of the wall-droplet gap width on the velocity. The gap increased with increasing the velocity, and the droplet elongated while distancing itself from the wall

6.2.3 *The Fahraeus Effect*

The Fahraeus effect describes a blood-related phenomenon which occurs in small tubes. It was proven experimentally by Fahraeus that the tube hematocrit HT is less than the discharge hematocrit HD. This happens when the RBC mean velocity V_{rbc} is higher than the mean blood velocity $\overline{V}$. The Fahraeus effect is described by the following equation (Sun and Munn 2005):

$$\frac{H_T}{H_D} = \frac{\overline{V}}{V_{rbc}} \tag{6.10}$$

Results from the simulations of Sect. 6.2.2 were revisited to analyze the effect of the RBC velocity on the tube hematocrit and were shown in Fig. 6.9. A comparison

Fig. 6.9 Decrease in tube hematocrit with respect to discharged hematocrit due to an increase in RBC velocity, and a comparison with Secomb results (Secomb et al. 2001)

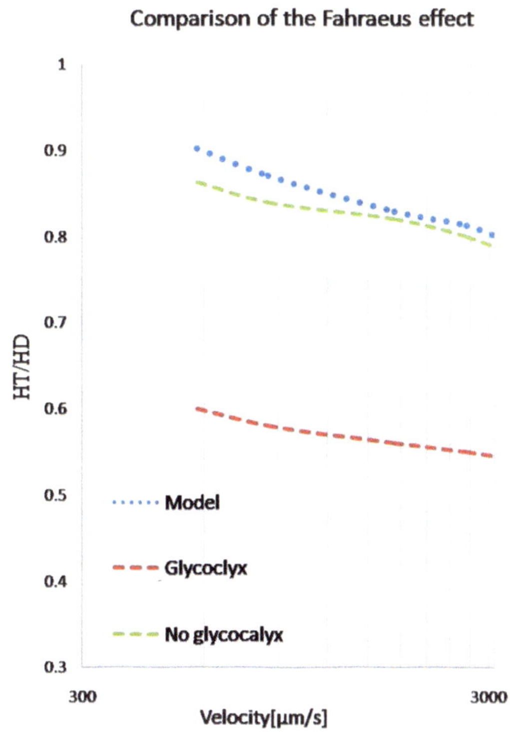

with the results of Secomb et al. (2001) showed that the model output was closer to Secomb's no glycocalyx model, which disregarded the endothelial surface layer effects. The reason for this behavior could be due to the sensitivity of the present model to the near-wall viscous effects. This resulted in a relatively lower droplet mean velocity, hence a higher value of the Fahraeus effect. With the increase in velocity, the droplet was excluded from the wall. This reduced the shear stresses and the value of the Fahraeus effect.

To check the influence of varying the blood hematocrit in a narrow vessel on the Fahraeus effect, a 285×41 lattice sites were used for three runs with different number of droplets placed in the domain as shown in Fig. 6.10. This was to simulate 20 μm diameter tube experiment, whose results were presented by Sun and Munn (2005).

Periodic conditions were applied at the inlet and outlet boundaries and mixed boundary condition at the top and bottom walls. This is a special boundary condition consisting of a combination of the full bounce-back, and the specular-reflection boundary condition. More details on this boundary condition are provided by Shirani and Jafari (2007). The mixed boundary condition was used because the results of Sun and Munn (2005) were compared with in vitro experimental data. The results of the three simulations which were performed with discharge hematocrit 0.17, 0.25, and 0.33, respectively, are presented in Fig. 6.11. The average Reynolds

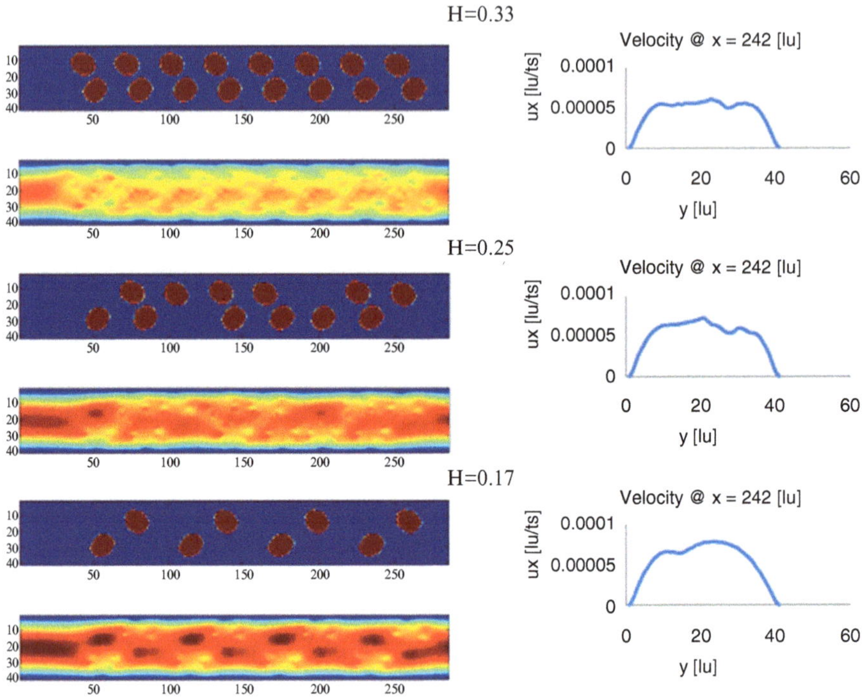

Fig. 6.10 Left-phase field contours and horizontal velocity profile for 285×41 lattice sites and 0.17, 0.25, and 0.33 discharge hematocrit. Right-horizontal velocity profile measured at the middle of the domain

number was $Re \approx 0.025$ which is a representative value for blood flow in the venules. The mean droplet velocity was calculated by Eq. (6.8). The Fahraeus effect was evaluated using Eq. (6.10). A reflection coefficient of 0.7 was used for this simulation.

6.2.4 *The Fahraeus–Lindqvist Effect*

The Fahraeus–Lindqvist effect predicts a decrease in the apparent viscosity of blood in a long narrow vessel of diameter ranging from 7 to 200 μm. This is due to the presence of a cell-free layer, referred to as plasma-skimming layer near the wall. In the absence of gravitational effects and under shear or parabolic flow it is known that when neutrally buoyant droplets (Legendre and Magnaudet 1998) or vesicles (Seifert 1999) are in proximity to a channel walls, they tend to migrate towards the center due to a hydrodynamic shear lift.

With 16 lattice units for the RBC diameter, a 57×240 lattice squares domain was used to investigate the model capability of reproducing a plasma-skimming layer

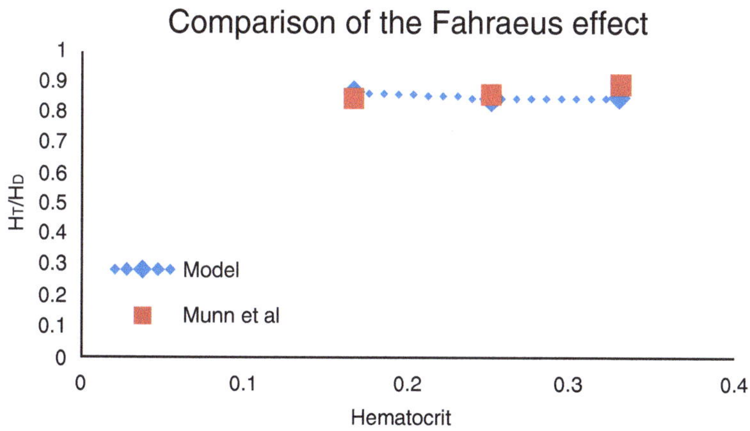

Fig. 6.11 The graph represents the dependence of the Fahraeus effect on the discharge hematocrit. Comparison of the model results with the results of Sun and Munn (2005)

based on the shear lift phenomenon. Eight out of 12 droplets were put initially on the walls. A Reynolds number Re $\approx$ 0.05 was used for this simulation. Periodic conditions were applied at the inlet and outlet boundaries to create a resemblance of a long tube and bounce-back at the upper and lower walls to impose zero velocity on the wall as described in the theory (Chandran et al. 2006).

This simulation clearly demonstrated that the near-wall droplets tilted due to the viscous effects of the surrounding fluid, and migrated towards the center as shown in the phase field contour insets of Fig. 6.12 (see Movie C6.2.4—Fahraeus). The pressure contour insets of Fig. 6.12 show a consistent pressure difference in the regions below and the regions above the droplets. This pressure difference, the droplet tilt, and its elongation are the major contributors to the creation of the lift force.

To trace the near-wall droplets mass center displacement in the vertical direction due to the shear lift, the normalized y coordinate by the channel width Y/H was plotted versus dimensionless time steps which were defined as $t\dot{\gamma}$ where t was the time in lattice units, and for convenience the shear strain rate was taken with respect to the height $H/4$ since in parabolic flows this rate varies with the coordinate of the location of interest in the vertical direction. This led to the following equation for the shear strain rate:

$$\dot{\gamma} = \frac{2U_0}{H} \tag{6.11}$$

where U_0 is the undisturbed centerline fluid input velocity. The droplets reached an equilibrium position at 21 % of the channel height. The result presented in Fig. 6.13 is indicative of the joint influence of the shear lift and the effect of the higher velocity droplets placed in the center of the channel, which halted the axial migration when the droplets came into close proximity.

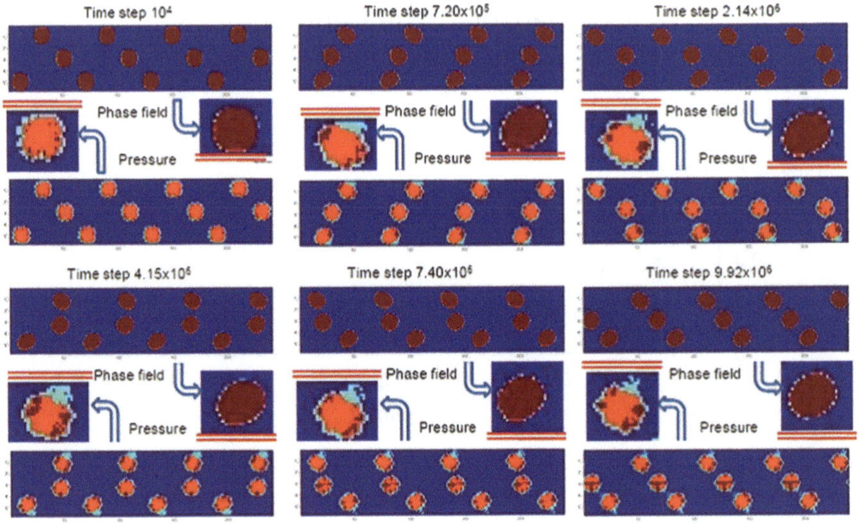

Fig. 6.12 Phase field contours and pressure contours for six different time steps. The pressure contours show a pressure difference between regions below and above the wall-side droplets. The phase field contours show the axial migration of the near-wall droplets

Fig. 6.13 Graph representing the axial migration of the near-walls droplets. Normalized displacement of the droplet mass center versus the corresponding time step multiplied by the shear rate of the mid location between the wall and the center of the channel

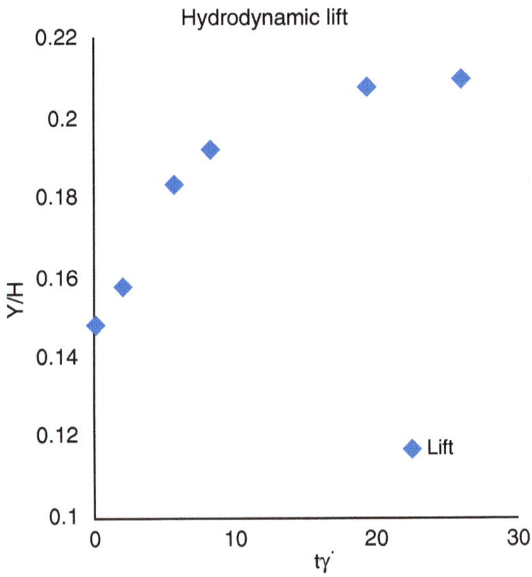

Appendix
Incorporating the Surfactant Effects into the Gunstensen LBM

The surfactant time-dependent convection-diffusion equation is given by the following:

$$\frac{\partial \Gamma}{\partial t} + \boldsymbol{\nabla}_s.(\boldsymbol{u}_s\Gamma) + k\Gamma u_n = D_s \boldsymbol{\nabla}_s^2 \Gamma + q_{chem} + q_f \tag{A1}$$

For insoluble, nondiffusing surfactant convected on the interface by the flow only and in the absence of chemical reaction, the time-dependent convection-diffusion equation takes the following form:

$$\frac{\partial \Gamma}{\partial t} + \nabla_s.(\boldsymbol{u}_s\Gamma) + k\Gamma u_n = 0 \tag{A2}$$

For 2D domain

The local mean curvature is calculated by $k = \boldsymbol{\nabla}_s \cdot n$

$$\begin{aligned}
k &= \nabla_s.n = (I - nn).\nabla.n = \nabla.n - n(n.\nabla).n \\
&= \frac{\partial n_x}{\partial x} + \frac{\partial n_y}{\partial y} - (n_x i + n_y j)\left(n_x \frac{\partial}{\partial x} + n_y \frac{\partial}{\partial y}\right).(n_x i + n_y j) \\
&= \frac{\partial n_x}{\partial x} + \frac{\partial n_y}{\partial y} - \left[(n_x i + n_y j).\left(n_x \frac{\partial n_x}{\partial x} i + n_x \frac{\partial n_y}{\partial x} j + n_y \frac{\partial n_x}{\partial y} i + n_y \frac{\partial n_y}{\partial y} j\right)\right] \\
&= \frac{\partial n_x}{\partial x} + \frac{\partial n_y}{\partial y} - \left(n_x^2 \frac{\partial n_x}{\partial x} + n_x n_y \frac{\partial n_x}{\partial y} + n_x n_y \frac{\partial n_y}{\partial x} + n_y^2 \frac{\partial n_y}{\partial y}\right) \\
&= \frac{\partial n_x}{\partial x}\left(1 - n_x^2\right) + \frac{\partial n_y}{\partial y}\left(1 - n_y^2\right) - n_x n_y\left(\frac{\partial n_x}{\partial y} + \frac{\partial n_y}{\partial x}\right) \\
&= n_y^2 \frac{\partial n_x}{\partial x} + n_x^2 \frac{\partial n_y}{\partial y} - n_x n_y\left(\frac{\partial n_x}{\partial y} + \frac{\partial n_y}{\partial x}\right)
\end{aligned} \tag{A3}$$

H. Farhat et al., *Accelerated Lattice Boltzmann Model for Colloidal Suspensions: Rheology and Interface Morphology*, DOI 10.1007/978-1-4899-7402-0,

The normal velocity is given by $u_n = \boldsymbol{u} \cdot \boldsymbol{n}$

$$u_n = u_x n_x + u_y n_y \tag{A4}$$

The tangential velocity is calculated by $\boldsymbol{u}_s = (\boldsymbol{I} - \boldsymbol{nn}) \cdot \boldsymbol{u}$

$$\begin{aligned} u_s &= (I - nn).u = u - n(n.u) = \left(u_x i + u_y j\right) - \left(n_x i + n_y j\right)\left(n_x u_x + n_y u_y\right) \\ &= \left(u_x - n_x^2 u_x - n_x n_y u_y\right) i + \left(u_y - n_y^2 u_y - n_x n_y u_x\right) j = u_{sx} i + u_{sy} j \end{aligned} \tag{A5}$$

The term $\nabla_s(\boldsymbol{u}_s\Gamma)$ is resolved by the product rule as follows:

$$\nabla_s.(u_s\Gamma) = \Gamma(\nabla_s.u_s) + u_s.\nabla_s\Gamma \tag{A6}$$

$$\begin{aligned} \nabla_s\Gamma &= \nabla\Gamma - n(n.\nabla\Gamma) \\ &= \left[\left(\frac{\partial\Gamma}{\partial x} i + \frac{\partial\Gamma}{\partial y} j\right) - \left(n_x i + n_y j\right)\left(n_x \frac{\partial\Gamma}{\partial x} + n_y \frac{\partial\Gamma}{\partial y}\right)\right] \\ &= \left[\left(\frac{\partial\Gamma}{\partial x} i + \frac{\partial\Gamma}{\partial y} j\right) - \left(n_x^2 \frac{\partial\Gamma}{\partial x} i + n_x n_y \frac{\partial\Gamma}{\partial y} i + n_x n_y \frac{\partial\Gamma}{\partial x} j + n_y^2 \frac{\partial\Gamma}{\partial y} i\right)\right] \\ &= \left(n_y^2 \frac{\partial\Gamma}{\partial x} - n_x n_y \frac{\partial\Gamma}{\partial y}\right) i + \left(n_x^2 \frac{\partial\Gamma}{\partial y} - n_x n_y \frac{\partial\Gamma}{\partial x}\right) j \end{aligned} \tag{A7}$$

$$\begin{aligned} u_s.\nabla_s\Gamma &= \left(u_{sx} i + u_{sy} j\right).\left\{\left(\left[1 - n_x^2\right]\frac{\partial\Gamma}{\partial x} - n_x n_y \frac{\partial\Gamma}{\partial y}\right) i + \left(\left[1 - n_y^2\right]\frac{\partial\Gamma}{\partial y} - n_x n_y \frac{\partial\Gamma}{\partial x}\right) j\right\} \\ u_s.\nabla_s\Gamma &= \left(u_{sx} - n_x^2 u_{sx} - n_x n_y u_{sy}\right)\frac{\partial\Gamma}{\partial x} + \left(u_{sy} - n_y^2 u_{sy} - n_x n_y u_{sx}\right)\frac{\partial\Gamma}{\partial y} = u_{sx}\frac{\partial\Gamma}{\partial x} + u_{sy}\frac{\partial\Gamma}{\partial y} \end{aligned} \tag{A8}$$

$$\begin{aligned} n_x^2 u_{sx} + n_x n_y u_{sy} &= n_x^2\left(u_x - n_x^2 u_x - n_x n_y u_y\right) + n_x n_y\left(u_y - n_y^2 u_y - n_x n_y u_x\right) \\ &= n_x^2 u_x - n_x^4 u_x - n_x^3 n_y u_y + n_x n_y u_y - n_x n_y^3 u_y - n_x^2 n_y^2 u_x n_x^2 u_{sx} + n_x n_y u_{sy} \\ &= \left[n_x^2 u_x - n_x^2\left(1 - n_y^2\right) u_x - n_x^2 n_y^2 u_x\right] + n_x n_y\left(u_y - n_x^2 u_y - n_y^2 u_y\right) = 0 \\ n_y^2 u_{sy} - n_x n_y u_{sx} &= n_y^2\left(u_y - n_y^2 u_y - n_x n_y u_x\right) + n_x n_y\left(u_x - n_x^2 u_x - n_x n_y u_y\right) \\ &= n_y^2 u_y - n_y^4 u_y - n_x n_y^3 u_x + n_x n_y u_x - n_x^3 n_y u_x - n_x^2 n_y^2 u_y n_y^2 u_{sy} + n_x n_y u_{sx} \\ &= \left[n_y^2 u_y - n_y^2(1 - n_x^2) u_y - n_x^2 n_y^2 u_y\right] + n_x n_y\left(u_x - n_y^2 u_x - n_x^2 u_x\right) = 0 \end{aligned}$$

$$
\begin{aligned}
\nabla_s.u_s &= \nabla.u_s - n(n.\nabla).u_s \\
&= \frac{\partial u_{sx}}{\partial x} + \frac{\partial u_{sy}}{\partial y} - (n_x i + n_y j)\left(n_x \frac{\partial}{\partial x} + n_y \frac{\partial}{\partial y}\right).(u_{sx} i + u_{sy} j) \\
&= \frac{\partial u_{sx}}{\partial x} + \frac{\partial u_{sy}}{\partial y} - \left(n_x^2 \frac{\partial}{\partial x} i + n_x n_y \frac{\partial}{\partial y} + n_x n_y \frac{\partial}{\partial x} j + n_y^2 \frac{\partial}{\partial y} j\right).(u_{sx} i + u_{sy} j) \\
&= \frac{\partial u_{sx}}{\partial x} + \frac{\partial u_{sy}}{\partial y} - n_x^2 \frac{\partial u_{sx}}{\partial x} - n_x n_y \frac{\partial u_{sx}}{\partial y} - n_y^2 \frac{\partial u_{sy}}{\partial y} - n_x n_y \frac{\partial u_{sy}}{\partial x} \\
&= n_y^2 \frac{\partial u_{sy}}{\partial y} + n_x^2 \frac{\partial u_{sx}}{\partial x} - n_x n_y \left(\frac{\partial u_{sx}}{\partial y} + \frac{\partial u_{sy}}{\partial x}\right)
\end{aligned} \tag{A9}
$$

$$
\nabla_s.(u_s \Gamma) = \Gamma\left[n_y^2 \frac{\partial u_{sx}}{\partial x} + n_x^2 \frac{\partial u_{sy}}{\partial y} - n_x n_y \left(\frac{\partial u_{sx}}{\partial y} + \frac{\partial u_{sy}}{\partial x}\right)\right] + u_{sx} \frac{\partial \Gamma}{\partial x} + u_{sy} \frac{\partial \Gamma}{\partial y} \tag{A10}
$$

The final form of the surfactant convection-diffusion equation is given by

$$
\begin{aligned}
\frac{\partial \Gamma}{\partial t} + \nabla_s.(u_s \Gamma) + k\Gamma u_n = {} & \frac{\partial \Gamma}{\partial t} + u_{sx} \frac{\partial \Gamma}{\partial x} + u_{sy} \frac{\partial \Gamma}{\partial y} \\
& + \Gamma\left[n_y^2 \frac{\partial u_{sx}}{\partial x} + n_x^2 \frac{\partial u_{sy}}{\partial y} - n_x n_y \left(\frac{\partial u_{sx}}{\partial y} + \frac{\partial u_{sy}}{\partial x}\right)\right. \\
& \left. + (k u_x n_x + k u_y n_y)\right] = 0
\end{aligned} \tag{A11}
$$

In short notation the convection-diffusion equation reads as

$$
\frac{\partial \Gamma}{\partial t} + C1 \frac{\partial \Gamma}{\partial x} + C2 \frac{\partial \Gamma}{\partial y} + C3\Gamma = 0 \tag{A12}
$$

where the coefficients are calculated by

$$
\begin{aligned}
C_1 &= u_{sx} \\
C_2 &= u_{sy}
\end{aligned} \tag{A13}
$$

$$
C_3 = n_y^2 \frac{\partial u_{sx}}{\partial x} + n_x^2 \frac{\partial u_{sy}}{\partial y} - n_x n_y \left(\frac{\partial u_{sx}}{\partial y} + \frac{\partial u_{sy}}{\partial x}\right) + (k u_x n_x + k u_y n_y)
$$

The explicit part of the hopscotch scheme is given by

$$
\Gamma_{i,j}^{n+1} = \frac{C_1}{2}\left(\Gamma_{i-1,j}^{n} - \Gamma_{i+1,j}^{n}\right) + \frac{C_2}{2}\left(\Gamma_{i,j-1}^{n} - \Gamma_{i,j+1}^{n}\right) + (1 - C_3)\Gamma_{i,j}^{n} \tag{A14}
$$

The implicit part of hopscotch scheme is given by

$$\Gamma_{i,j}^{n+1} = \frac{1}{(1+C_3)}\left(\Gamma_{i,j}^{n} + \frac{C_1}{2}\left[\Gamma_{i-1,j}^{n+1} - \Gamma_{i+1,j}^{n+1}\right] + \frac{C_2}{2}\left[\Gamma_{i,j-1}^{n+1} - \Gamma_{i,j+1}^{n+1}\right]\right) \tag{A15}$$

When diffusion is to be considered the following equation should be added to the right-hand side of the time-dependent convection-diffusion equation:

$$\nabla_s \Gamma = \left(n_y^2 \frac{\partial \Gamma}{\partial x} - n_x n_y \frac{\partial \Gamma}{\partial y}\right) i + \left(n_x^2 \frac{\partial \Gamma}{\partial y} - n_x n_y \frac{\partial \Gamma}{\partial x}\right) j \tag{A16}$$

$$\begin{aligned}
\nabla_s \Gamma &= \nabla_s . \nabla_s \Gamma = \nabla_s . \left[\left(n_y^2 \frac{\partial \Gamma}{\partial x} - n_x n_y \frac{\partial \Gamma}{\partial y}\right) i + \left(n_x^2 \frac{\partial \Gamma}{\partial y} - n_x n_y \frac{\partial \Gamma}{\partial x}\right) j\right] \\
\nabla_s^2 \Gamma &= n_y^2 \frac{\partial}{\partial x}\left(n_y^2 \frac{\partial \Gamma}{\partial x} - n_x n_y \frac{\partial \Gamma}{\partial y}\right) + n_x^2 \frac{\partial}{\partial y}\left(n_y^2 \frac{\partial \Gamma}{\partial y} - n_x n_y \frac{\partial \Gamma}{\partial x}\right) \\
&\quad - n_x n_y \left[\frac{\partial}{\partial y}\left(n_y^2 \frac{\partial \Gamma}{\partial x} - n_x n_y \frac{\partial \Gamma}{\partial y}\right) + \frac{\partial}{\partial x}\left(n_y^2 \frac{\partial \Gamma}{\partial x} - n_x n_y \frac{\partial \Gamma}{\partial y}\right)\right] \\
\nabla_s^2 \Gamma &= n_y^4 \frac{\partial^2 \Gamma}{\partial x^2} - n_x n_y^3 \frac{\partial \Gamma}{\partial x \partial y} + n_x^4 \frac{\partial^2 \Gamma}{\partial y^2} - n_x^3 n_y \frac{\partial^2 \Gamma}{\partial x \partial y} - n_x n_y^3 \frac{\partial^2 \Gamma}{\partial x \partial y} + n_x^2 n_y^2 \frac{\partial^2 \Gamma}{\partial y^2} \\
&\quad - n_x^3 n_y \frac{\partial^2 \Gamma}{\partial x \partial y} + n_x^2 n_y^2 \frac{\partial^2 \Gamma}{\partial x^2} \\
\nabla_s^2 \Gamma &= \left(n_y^4 + n_x^2 n_y^2\right) \frac{\partial^2 \Gamma}{\partial x^2} + \left(n_x^4 + n_x^2 n_y^2\right) \frac{\partial^2 \Gamma}{\partial y^2} - 2\left(n_x^3 n_y + n_x n_y^3\right) \frac{\partial^2 \Gamma}{\partial x \partial y} \\
\nabla_s^2 \Gamma &= n_y^2\left(n_y^2 + n_x^2\right) \frac{\partial^2 \Gamma}{\partial x^2} + n_x^2\left(n_x^2 + n_y^2\right) \frac{\partial^2 \Gamma}{\partial y^2} - 2 n_x n_y \left(n_x^2 + n_y^2\right) \frac{\partial^2 \Gamma}{\partial x \partial y} \\
\nabla_s^2 \Gamma &= n_y^2 \frac{\partial^2 \Gamma}{\partial x^2} + n_x^2 \frac{\partial^2 \Gamma}{\partial y^2} - 2 n_x n_y \frac{\partial^2 \Gamma}{\partial x \partial y}
\end{aligned} \tag{A17}$$

In short notation the convection-diffusion equation reads as follows:

$$\frac{\partial \Gamma}{\partial t} + C_1 \frac{\partial \Gamma}{\partial x} + C_2 \frac{\partial \Gamma}{\partial y} + C_3 \Gamma + C_4 \frac{\partial^2 \Gamma}{\partial x^2} + C_5 \frac{\partial^2 \Gamma}{\partial y^2} + C_6 \frac{\partial^2 \Gamma}{\partial x \partial y} = 0 \tag{A18}$$

$$\begin{aligned}
C_4 &= -n_y^2 Ds = \left(n_x^2 - 1\right) Ds \\
C_5 &= -n_x^2 Ds = \left(n_y^2 - 1\right) Ds \\
C_6 &= 2 n_x n_y Ds
\end{aligned} \tag{A19}$$

The explicit equation is as follows:

$$\begin{aligned}\Gamma_{i,j}^{n+1} = &\frac{C_1}{2}\left(\Gamma_{i-1,j}^{n} - \Gamma_{i+1,j}^{n}\right) + \frac{C_2}{2}\left(\Gamma_{i,j-1}^{n} - \Gamma_{i,j+1}^{n}\right)\\ &+ (1 - C_3 + 2C_4 + 2C_5)\Gamma_{i,j}^{n} - C_4\left(\Gamma_{i+1,j}^{n} - \Gamma_{i-1,j}^{n}\right)\\ &- C_5\left(\Gamma_{i,j+1}^{n} + \Gamma_{i,j-1}^{n}\right)\\ &+ \frac{C_6}{2}\left(\Gamma_{i,j-1}^{n} - \Gamma_{i,j+1}^{n} + \Gamma_{i-1,j+1}^{n} - \Gamma_{i-1,j-1}^{n}\right)\end{aligned} \tag{A20}$$

The implicit equation is as follows:

$$\begin{aligned}\Gamma_{i,j}^{n+1} = &\frac{1}{(1 + C_3 - 2C_4 - 2C_5)}\left(\Gamma_{i,j}^{n} + \frac{C_1}{2}\left[\Gamma_{i-1,j}^{n+1} - \Gamma_{i+1,j}^{n+1}\right]\right.\\ &+ \frac{C_2}{2}\left[\Gamma_{i,j-1}^{n+1} - \Gamma_{i,j+1}^{n+1}\right] - C_4\left[\Gamma_{i+1,j}^{n+1} - \Gamma_{i-1,j}^{n+1}\right] - C_5\left[\Gamma_{i,j+1}^{n+1} - \Gamma_{i,j-1}^{n+1}\right]\\ &\left.+ \frac{C_6}{2}\left[\Gamma_{i,j-1}^{n+1} - \Gamma_{i,j+1}^{n+1} + \Gamma_{i-1,j+1}^{n+1} - \Gamma_{i-1,j-1}^{n+1}\right]\right)\end{aligned} \tag{A21}$$

The interfacial tension surface gradient is calculated as follows:

$$\begin{aligned}\nabla_s \sigma &= \left[\left(\frac{\partial \sigma}{\partial x} i + \frac{\partial \sigma}{\partial y} j\right) - (n_x i + n_y j)\left(n_x \frac{\partial \sigma}{\partial x} + n_y \frac{\partial \sigma}{\partial y}\right)\right]\\ &= \left[\left(\frac{\partial \sigma}{\partial x} i + \frac{\partial \sigma}{\partial y} j\right) - \left(n_x^2 \frac{\partial \sigma}{\partial x} i + n_x n_y \frac{\partial \sigma}{\partial y} i + n_x n_y \frac{\partial \sigma}{\partial x} j + n_y^2 \frac{\partial \sigma}{\partial y} j\right)\right]\\ &= \left(\frac{\partial \sigma}{\partial x} - n_x^2 \frac{\partial \sigma}{\partial x} - n_x n_y \frac{\partial \sigma}{\partial y}\right) i + \left(\frac{\partial \sigma}{\partial y} - n_y^2 \frac{\partial \sigma}{\partial y} - n_x n_y \frac{\partial \sigma}{\partial x}\right) j\end{aligned} \tag{A22}$$

For 3D domain

$$k=\nabla_s.n=(I-nn).\nabla.n=\nabla.n-n\left(n.\nabla\right).n=\frac{\partial n_x}{\partial x}+\frac{\partial n_y}{\partial y}+\frac{\partial n_y}{\partial y}-(n_xi+n_yj+n_zk)\left(n_x\frac{\partial}{\partial x}+n_y\frac{\partial}{\partial y}+n_z\frac{\partial}{\partial z}\right).(n_xi+n_yj+n_zk)$$

$$k=\frac{\partial n_x}{\partial x}+\frac{\partial n_y}{\partial y}+\frac{\partial n_z}{\partial z}-\left[(n_xi+n_yj+n_zk).\left(n_x\frac{\partial n_x}{\partial x}i+n_x\frac{\partial n_y}{\partial x}j+n_x\frac{\partial n_z}{\partial x}k+n_y\frac{\partial n_x}{\partial y}i+n_y\frac{\partial n_y}{\partial y}j+n_y\frac{\partial n_z}{\partial y}k+n_z\frac{\partial n_x}{\partial z}i+n_z\frac{\partial n_y}{\partial z}j+n_z\frac{\partial n_z}{\partial z}k\right)\right]$$

$$k=\frac{\partial n_x}{\partial x}+\frac{\partial n_y}{\partial y}+\frac{\partial n_z}{\partial z}-\left(n_x^2\frac{\partial n_x}{\partial x}+n_xn_y\frac{\partial n_x}{\partial y}+n_xn_z\frac{\partial n_x}{\partial z}+n_xn_y\frac{\partial n_y}{\partial x}+n_y^2\frac{\partial n_y}{\partial y}+n_yn_z\frac{\partial n_y}{\partial z}+n_xn_z\frac{\partial n_z}{\partial x}+n_yn_z\frac{\partial n_z}{\partial y}+n_z^2\frac{\partial n_z}{\partial z}\right)$$

$$k=\frac{\partial n_x}{\partial x}(1-n_x^2)+\frac{\partial n_y}{\partial y}\left(1-n_y^2\right)+\frac{\partial n_z}{\partial z}(1-n_z^2)-n_xn_y\left(\frac{\partial n_x}{\partial y}+\frac{\partial n_y}{\partial x}\right)-n_xn_z\left(\frac{\partial n_x}{\partial z}+\frac{\partial n_z}{\partial x}\right)-n_yn_z\left(\frac{\partial n_y}{\partial z}+\frac{\partial n_z}{\partial y}\right)$$

$$k=\left(n_y^2+n_z^2\right)\frac{\partial n_x}{\partial x}+(n_x^2+n_z^2)\frac{\partial n_y}{\partial y}+\left(n_x^2+n_y^2\right)\frac{\partial n_z}{\partial z}-n_xn_y\left(\frac{\partial n_x}{\partial y}+\frac{\partial n_y}{\partial x}\right)-n_xn_z\left(\frac{\partial n_x}{\partial z}+\frac{\partial n_z}{\partial x}\right)-n_yn_z\left(\frac{\partial n_y}{\partial z}+\frac{\partial n_z}{\partial y}\right) \tag{A23}$$

The normal velocity is given by

$$u_n = u_x n_x + u_y n_y + u_z n_z \tag{A24}$$

The tangential velocity is given by

$$\begin{aligned} u_s &= (I - nn).u = u - n(n.u) = (u_x i + u_y j + u_z k) \\ &\quad - (n_x i + n_y j + n_z k)(n_x u_x + n_y u_y + n_z u_z) \\ &= (u_x - n_x^2 u_x - n_x n_y u_y - n_x n_z u_z) i + \left(u_y - n_y n_x u_x - n_y^2 u_y - n_y n_z u_z\right) j \\ &\quad + (u_z - n_z n_x u_x - n_z n_y u_y - n_z^2 u_z) k = u_{sx} i + u_{sy} j + u_{sz} k \end{aligned} \tag{A25}$$

By the product rule the following term is solved as follows:

$$\begin{aligned} &\nabla_s.(u_s \Gamma) = \Gamma(\nabla_s.u_s) + u_s.\nabla_s \Gamma \\ &\nabla_s \Gamma = \nabla \Gamma - n(n.\nabla \Gamma) = \left(\frac{\partial \Gamma}{\partial x} i + \frac{\partial \Gamma}{\partial y} j + \frac{\partial \Gamma}{\partial z} k\right) \\ &\qquad - (n_x i + n_y j + n_z k)\left(n_x \frac{\partial \Gamma}{\partial x} + n_y \frac{\partial \Gamma}{\partial y} + n_z \frac{\partial \Gamma}{\partial z}\right) \\ &\nabla_s \Gamma = \left(\frac{\partial \Gamma}{\partial x} i + \frac{\partial \Gamma}{\partial y} j + \frac{\partial \Gamma}{\partial z} k\right) - n_x^2 \frac{\partial \Gamma}{\partial x} i - n_x n_y \frac{\partial \Gamma}{\partial y} i - n_x n_z \frac{\partial \Gamma}{\partial z} i - n_x n_y \frac{\partial \Gamma}{\partial x} j - n_y^2 \frac{\partial \Gamma}{\partial y} j \\ &\qquad - n_y n_z \frac{\partial \Gamma}{\partial z} j - n_x n_z \frac{\partial \Gamma}{\partial x} k - n_y n_z \frac{\partial \Gamma}{\partial y} k - n_z^2 \frac{\partial \Gamma}{\partial z} k \\ &\nabla_s \Gamma = \left[(1 - n_x^2)\frac{\partial \Gamma}{\partial x} - n_x n_y \frac{\partial \Gamma}{\partial y} - n_x n_z \frac{\partial \Gamma}{\partial z}\right] i + \left[\left(1 - n_y^2\right)\frac{\partial \Gamma}{\partial y} - n_x n_y \frac{\partial \Gamma}{\partial x} - n_y n_z \frac{\partial \Gamma}{\partial z}\right] j \\ &\qquad + \left[(1 - n_z^2)\frac{\partial \Gamma}{\partial z} - n_x n_z \frac{\partial \Gamma}{\partial x} - n_y n_z \frac{\partial \Gamma}{\partial y}\right] k \end{aligned} \tag{A26}$$

$$\begin{aligned} &u_s.\nabla_s \Gamma = (u_{sx} i + u_{sy} j + u_{sz} j).\left\{\left[(1 - n_x^2)\frac{\partial \Gamma}{\partial x} - n_x n_y \frac{\partial \Gamma}{\partial y} - n_x n_z \frac{\partial \Gamma}{\partial z}\right] i\right. \\ &\left. + \left[\left(1 - n_y^2\right)\frac{\partial \Gamma}{\partial y} - n_x n_y \frac{\partial \Gamma}{\partial x} - n_y n_z \frac{\partial \Gamma}{\partial z}\right] j + \left[(1 - n_z^2)\frac{\partial \Gamma}{\partial z} - n_x n_z \frac{\partial \Gamma}{\partial x} - n_y n_z \frac{\partial \Gamma}{\partial y}\right] k\right\} \\ &u_s.\nabla_s \Gamma = (u_{sx} - n_x^2 u_{sx} - n_x n_y u_{sy} - n_x n_z u_{sz})\frac{\partial \Gamma}{\partial x} \\ &+ \left(u_{sy} - n_y n_x u_{sx} - n_y^2 u_{sy} - n_y n_z u_{sz}\right)\frac{\partial \Gamma}{\partial y} + (u_{sz} - n_z n_x u_{sx} - n_z n_y u_{sy} - n_z^2 u_{sz})\frac{\partial \Gamma}{\partial z} \end{aligned} \tag{A27}$$

$$
\begin{aligned}
&n_x^2 u_{sx} + n_x n_y u_{sy} + n_x n_z u_{sz} = n_x^2 u_x - n_x^4 u_x + n_x^3 n_y u_y - n_x^3 n_z u_z + n_x n_y u_y \\
&-n_x^2 n_y^2 u_x - n_x n_y^3 u_y - n_x n_y^2 n_z u_z + n_x n_z u_z - n_x^2 n_z^2 u_x - n_x n_y n_z^2 u_y - n_x n_z^3 u_z \\
&n_x^2 u_{sx} + n_x n_y u_{sy} + n_x n_z u_{sz} = \left[n_x^2 u_x - n_x^2\left(1 - n_y^2 - n_z^2\right)u_x - n_x^2 n_y^2 u_x - n_x^2 n_z^2 u_x\right] \\
&+n_x n_y\left(u_y - n_x^2 u_y - n_y^2 u_y - n_y^2 u_y\right) + n_x n_z\left(u_z - n_x^2 u_z - n_y^2 u_z - n_y^2 u_z\right) = 0 \\
&n_y n_x u_{sx} + n_y^2 u_{sy} + n_y n_z u_{sz} = 0 \\
&n_z n_x u_{sx} + n_z n_y u_{sy} + n_z^2 u_{sz} = 0 \\
&\nabla_s . u_s = \nabla . u_s - n(n.\nabla).u_s = \frac{\partial u_{sx}}{\partial x} + \frac{\partial u_{sy}}{\partial y} + \frac{\partial u_{sz}}{\partial z} - \left(n_x i + n_y j + n_z k\right) \\
&\left(n_x \frac{\partial}{\partial x} + n_y \frac{\partial}{\partial y} + n_z \frac{\partial}{\partial z}\right).\left(u_{sx} i + u_{sy} j + u_{sz} k\right) \\
&\nabla_s . u_s = \frac{\partial u_{sx}}{\partial x} + \frac{\partial u_{sy}}{\partial y} + \frac{\partial u_{sz}}{\partial z} - \left(n_x^2 \frac{\partial}{\partial x} i + n_x n_y \frac{\partial}{\partial y} i + n_x n_z \frac{\partial}{\partial z} i + n_y n_x \frac{\partial}{\partial x} j\right. \\
&\left. + n_y^2 \frac{\partial}{\partial y} j + n_y n_z \frac{\partial}{\partial z} j + n_z n_x \frac{\partial}{\partial x} k + n_z n_y \frac{\partial}{\partial y} k + n_z^2 \frac{\partial}{\partial z} k\right).\left(u_{sx} i + u_{sy} j + u_{sz} k\right) \\
&\nabla_s . u_s = \frac{\partial u_{sx}}{\partial x} + \frac{\partial u_{sy}}{\partial y} + \frac{\partial u_{sz}}{\partial z} - n_x^2 \frac{\partial u_{sx}}{\partial x} - n_x n_y \frac{\partial u_{sx}}{\partial y} - n_x n_z \frac{\partial u_{sx}}{\partial z} - n_y n_x \frac{\partial u_{sy}}{\partial x} \\
&-n_y^2 \frac{\partial u_{sy}}{\partial y} - n_y n_z \frac{\partial u_{sy}}{\partial z} - n_z n_x \frac{\partial u_{sz}}{\partial x} - n_z n_y \frac{\partial u_{sz}}{\partial y} - n_z^2 \frac{\partial u_{sz}}{\partial z} \\
&\nabla_s . u_s = \left(n_y^2 + n_z^2\right)\frac{\partial u_{sx}}{\partial x} + \left(n_x^2 + n_z^2\right)\frac{\partial u_{sy}}{\partial y} + \left(n_x^2 + n_y^2\right)\frac{\partial u_{sz}}{\partial z} \\
&-n_x n_y\left(\frac{\partial u_{sx}}{\partial y} + \frac{\partial u_{sy}}{\partial x}\right) - n_x n_z\left(\frac{\partial u_{sx}}{\partial z} + \frac{\partial u_{sz}}{\partial x}\right) - n_y n_z\left(\frac{\partial u_{sy}}{\partial z} + \frac{\partial u_{sz}}{\partial y}\right)
\end{aligned}
\tag{A28}
$$

In short notation the convection-diffusion equation reads

$$
\frac{\partial \Gamma}{\partial t} + D_1 \frac{\partial \Gamma}{\partial x} + D_2 \frac{\partial \Gamma}{\partial y} + D_3 \frac{\partial \Gamma}{\partial z} + D_4 \Gamma = 0 \tag{A29}
$$

where the coefficients are given by

$$
\begin{aligned}
D_1 &= u_{sx} \\
D_2 &= u_{sy} \\
D_3 &= u_{sz}
\end{aligned}
\tag{A30}
$$

$$D_4 = \left(n_y^2 + n_z^2\right)\frac{\partial u_{sx}}{\partial x} + \left(n_x^2 + n_z^2\right)\frac{\partial u_{sy}}{\partial y} + \left(n_x^2 + n_y^2\right)\frac{\partial u_{sz}}{\partial z} - n_x n_y\left(\frac{\partial u_{sx}}{\partial y} + \frac{\partial u_{sy}}{\partial x}\right)$$
$$- n_x n_z\left(\frac{\partial u_{sx}}{\partial z} + \frac{\partial u_{sz}}{\partial x}\right) - n_y n_z\left(\frac{\partial u_{sy}}{\partial z} + \frac{\partial u_{sz}}{\partial y}\right) + \left(ku_x n_x + ku_y n_y + ku_z n_z\right)$$

The hopscotch finite difference scheme is given as follows:
Explicit part:

$$\Gamma_{i,j,k}^{n+1} = \frac{C_1}{2}\left(\Gamma_{i-1,j,k}^{n} - \Gamma_{i+1,j,k}^{n}\right) + \frac{C_2}{2}\left(\Gamma_{i,j-1,k}^{n} - \Gamma_{i,j+1,k}^{n}\right)$$
$$+ \frac{C_3}{2}\left(\Gamma_{i,j,k-1}^{n} - \Gamma_{i,j,k+1}^{n}\right) + (1 - C_4)\Gamma_{i,j,k}^{n} \quad \text{(A31)}$$

Implicit part:

$$\Gamma_{i,j,k}^{n+1} = \frac{1}{(1 + D_4)}\left(\Gamma_{i,j,k}^{n} + \frac{D_1}{2}\left[\Gamma_{i-1,j,k}^{n+1} - \Gamma_{i+1,j,k}^{n+1}\right]\right.$$
$$\left. + \frac{D_2}{2}\left[\Gamma_{i,j-1,k}^{n+1} - \Gamma_{i,j+1,k}^{n+1}\right] + \frac{D_3}{2}\left[\Gamma_{i,j,k-1}^{n+1} - \Gamma_{i,j,k+1}^{n+1}\right]\right) \quad \text{(A32)}$$

The diffusion term is solved as follows:

$$
\begin{aligned}
\nabla_s^2\Gamma &= \nabla_s.\nabla_s\Gamma = \nabla_s.\left[(1-n_x^2)\frac{\partial\Gamma}{\partial x} - n_xn_y\frac{\partial\Gamma}{\partial y} - n_xn_z\frac{\partial\Gamma}{\partial z}\right]i + \left[\left(1-n_y^2\right)\frac{\partial\Gamma}{\partial y} - n_xn_y\frac{\partial\Gamma}{\partial x} - n_yn_z\frac{\partial\Gamma}{\partial z}\right]j + \left[(1-n_z^2)\frac{\partial\Gamma}{\partial z} - n_xn_z\frac{\partial\Gamma}{\partial x} - n_yn_z\frac{\partial\Gamma}{\partial y}\right]k \\
\nabla_s^2\Gamma &= (1-n_x^2)\frac{\partial}{\partial x}\left[(1-n_x^2)\frac{\partial\Gamma}{\partial x} - n_xn_y\frac{\partial\Gamma}{\partial y} - n_xn_z\frac{\partial\Gamma}{\partial z}\right] + \left(1-n_y^2\right)\frac{\partial}{\partial y}\left[\left(1-n_y^2\right)\frac{\partial\Gamma}{\partial y} - n_xn_y\frac{\partial\Gamma}{\partial x} - n_yn_z\frac{\partial\Gamma}{\partial z}\right] \\
&+(1-n_z^2)\frac{\partial}{\partial z}\left[(1-n_z^2)\frac{\partial\Gamma}{\partial z} - n_xn_z\frac{\partial\Gamma}{\partial x} - n_yn_z\frac{\partial\Gamma}{\partial y}\right] - n_xn_y\left(\frac{\partial}{\partial y}\left[(1-n_x^2)\frac{\partial\Gamma}{\partial x} - n_xn_y\frac{\partial\Gamma}{\partial y} - n_xn_z\frac{\partial\Gamma}{\partial z}\right]\right. \\
&\left.+\frac{\partial}{\partial x}\left[\left(1-n_y^2\right)\frac{\partial\Gamma}{\partial y} - n_xn_y\frac{\partial\Gamma}{\partial x} - n_yn_z\frac{\partial\Gamma}{\partial z}\right]\right) - n_xn_z\left(\frac{\partial}{\partial z}\left[(1-n_x^2)\frac{\partial\Gamma}{\partial x} - n_xn_y\frac{\partial\Gamma}{\partial y} - n_xn_z\frac{\partial\Gamma}{\partial z}\right]\right. \\
&\left.+\frac{\partial}{\partial x}\left[(1-n_z^2)\frac{\partial\Gamma}{\partial z} - n_xn_z\frac{\partial\Gamma}{\partial x} - n_yn_z\frac{\partial\Gamma}{\partial y}\right]\right) - n_yn_z\left(\frac{\partial}{\partial z}\left[\left(1-n_y^2\right)\frac{\partial\Gamma}{\partial y} - n_xn_y\frac{\partial\Gamma}{\partial x} - n_yn_z\frac{\partial\Gamma}{\partial z}\right] + \frac{\partial}{\partial y}\left[(1-n_z^2)\frac{\partial\Gamma}{\partial z} - n_xn_z\frac{\partial\Gamma}{\partial x} - n_yn_z\frac{\partial\Gamma}{\partial y}\right]\right) \\
\nabla_s^2\Gamma &= \left[(1-n_x^2)^2 + n_x^2n_y^2 + n_x^2n_z^2\right]\frac{\partial^2\Gamma}{\partial x^2} + \left[\left(1-n_y^2\right)^2 + n_x^2n_y^2 + n_y^2n_z^2\right]\frac{\partial^2\Gamma}{\partial y^2} + \left[(1-n_z^2)^2 + n_x^2n_z^2 + n_y^2n_z^2\right]\frac{\partial^2\Gamma}{\partial z^2} \\
&+2\left[n_xn_yn_z^2 - (1-n_x^2)n_xn_y - \left(1-n_y^2\right)n_xn_y\right]\frac{\partial^2\Gamma}{\partial x\partial y} + 2\left[n_xn_y^2n_z - (1-n_x^2)n_xn_z - (1-n_z^2)n_xn_z\right]\frac{\partial^2\Gamma}{\partial x\partial z} + 2\left[n_x^2n_yn_z - \left(1-n_y^2\right)n_yn_z - (1-n_z^2)n_yn_z\right]\frac{\partial^2\Gamma}{\partial y\partial z}
\end{aligned}
\tag{A33}
$$

$$\begin{aligned}\left(1-n_x^2\right)^2+n_x^2n_y^2+n_x^2n_z^2&=1-2n_x^2+n_x^4+n_x^2n_y^2+n_x^2n_z^2\\&=n_y^2+n_z^2-n_x^2+n_x^2\left(1-n_y^2-n_z^2\right)+n_x^2n_y^2+n_x^2n_z^2\\&=n_y^2+n_z^2=1-n_x^2\end{aligned}$$

$$\begin{aligned}\left(1-n_y^2\right)^2+n_x^2n_y^2+n_y^2n_z^2&=1-2n_y^2+n_y^4+n_x^2n_y^2+n_y^2n_z^2\\&=n_x^2+n_z^2-n_y^2+n_y^2\left(1-n_x^2-n_z^2\right)+n_x^2n_y^2+n_y^2n_z^2\\&=n_x^2+n_z^2=1-n_y^2\end{aligned}$$

$$\begin{aligned}\left(1-n_z^2\right)^2+n_x^2n_z^2+n_y^2n_z^2&=1-2n_z^2+n_z^4+n_x^2n_z^2+n_y^2n_z^2\\&=n_x^2+n_y^2-n_z^2+n_z^2\left(1-n_x^2-n_y^2\right)+n_x^2n_z^2+n_y^2n_z^2\\&=n_x^2+n_y^2=1-n_z^2\end{aligned}$$

$$2\left[n_xn_yn_z^2-\left(1-n_x^2\right)n_xn_y-\left(1-n_y^2\right)n_xn_y\right]=2\left[n_xn_y\left(n_z^2+n_y^2+n_z^2\right)-2n_xn_y\right]=-2n_xn_y$$

$$2\left[n_xn_y^2n_z-\left(1-n_x^2\right)n_xn_z-\left(1-n_z^2\right)n_xn_z\right]=2\left[n_xn_z\left(n_z^2+n_y^2+n_z^2\right)-2n_xn_z\right]=-2n_xn_z$$

$$2\left[n_x^2n_yn_z-\left(1-n_y^2\right)n_yn_z-\left(1-n_z^2\right)n_yn_z\right]=2\left[n_yn_z\left(n_z^2+n_y^2+n_z^2\right)-2n_yn_z\right]=-2n_yn_z$$

The equation for the time-dependent surfactant-convection equation is as follows:

$$\begin{aligned}&\frac{\partial\Gamma}{\partial t}+D1\frac{\partial\Gamma}{\partial x}+D2\frac{\partial\Gamma}{\partial y}+D3\frac{\partial\Gamma}{\partial z}+D4\Gamma+D5\frac{\partial^2\Gamma}{\partial x^2}+D6\frac{\partial^2\Gamma}{\partial y^2}+D7\frac{\partial^2\Gamma}{\partial z^2}\\&\quad+D8\frac{\partial^2\Gamma}{\partial x\partial y}+D9\frac{\partial^2\Gamma}{\partial x\partial z}+D10\frac{\partial^2\Gamma}{\partial y\partial z}=0\end{aligned}\tag{A34}$$

where the coefficients are

$$\begin{aligned}&D5=-\left(n_y{}^2+n_z{}^2\right)D_s=\left(n_x{}^2-1\right)D_s\\&D6=-\left(n_x{}^2+n_z{}^2\right)D_s=\left(n_y{}^2-1\right)D_s\\&D7=-\left(n_x{}^2+n_y{}^2\right)D_s=\left(n_z{}^2-1\right)D_s\\&D8=2n_xn_yD_s\\&D9=2n_xn_zD_s\\&D10=2n_yn_zD_s\end{aligned}\tag{A35}$$

The explicit part:

$$\begin{aligned}
\Gamma^{n+1}_{i,j,k} = & \frac{D_1}{2}\left(\Gamma^n_{i-1,j,k} - \Gamma^n_{i+1,j,k}\right) + \frac{D_2}{2}\left(\Gamma^n_{i,j-1,k} - \Gamma^n_{i,j+1,k}\right) \\
& + \frac{D_3}{2}\left(\Gamma^n_{i,j,k-1} - \Gamma^n_{i,j,k+1}\right) + (1 - D_4 + 2D_5 + 2D_6 + 2D_7)\Gamma^n_{i,j,k} \\
& - D_5\left(\Gamma^n_{i+1,j,k} + \Gamma^n_{i-1,j,k}\right) - D_6\left(\Gamma^n_{i,j+1,k} + \Gamma^n_{i,j-1,k}\right) \\
& - D_7\left(\Gamma^n_{i,j,k+1} + \Gamma^n_{i,j,k-1}\right) \\
& + \frac{D_8}{2}\left(\Gamma^n_{i,j-1,k} - \Gamma^n_{i,j+1,k} + \Gamma^n_{i-1,j+1,k} - \Gamma^n_{i-1,j-1,k}\right) \\
& + \frac{D_9}{2}\left(\Gamma^n_{i,j,k-1} - \Gamma^n_{i,j,k+1} + \Gamma^n_{i-1,j,k+1} - \Gamma^n_{i-1,j,k-1}\right) \\
& + \frac{D_{10}}{2}\left(\Gamma^n_{i,j,k-1} - \Gamma^n_{i,j,k+1} + \Gamma^n_{i,j-1,k+1} - \Gamma^n_{ij-1,,k-1}\right)
\end{aligned} \tag{A36}$$

The implicit part:

$$\begin{aligned}
\Gamma^{n+1}_{i,j,k} = & \frac{1}{(1 + D_4 - 2D_5 - 2D_6 - 2D_7)} \\
& \left(\Gamma^n_{i,j,k} + \frac{D_1}{2}\left[\Gamma^{n+1}_{i-1,j,k} - \Gamma^{n+1}_{i+1,j,k}\right] + \frac{D_2}{2}\left[\Gamma^{n+1}_{i,j-1,k} - \Gamma^{n+1}_{i,j+1,k}\right] + \frac{D_3}{2}\left[\Gamma^{n+1}_{i,j,k-1} - \Gamma^{n+1}_{i,j,k+1}\right]\right. \\
& - D_5\left[\Gamma^{n+1}_{i+1,j,k} + \Gamma^{n+1}_{i-1,j,k}\right] - D_6\left[\Gamma^{n+1}_{i,j+1,k} + \Gamma^{n+1}_{i,j-1,k}\right] - D_7\left[\Gamma^{n+1}_{i,j,k+1} + \Gamma^{n+1}_{i,j,k-1}\right] \\
& + \frac{D_8}{2}\left[\Gamma^{n+1}_{i,j-1,k} - \Gamma^{n+1}_{i,j+1,k} + \Gamma^{n+1}_{i-1,j+1,k} - \Gamma^{n+1}_{i-1,j-1,k}\right] \\
& + \frac{D_9}{2}\left[\Gamma^{n+1}_{i,j,k-1} - \Gamma^{n+1}_{i,j,k+1} + \Gamma^{n+1}_{i-1,j,k+1} - \Gamma^{n+1}_{i-1,j,k-1}\right] \\
& \left. + \frac{D_{10}}{2}\left[\Gamma^{n+1}_{i,j,k-1} - \Gamma^{n+1}_{i,j,k+1} + \Gamma^{n+1}_{i,j-1,k+1} - \Gamma^{n+1}_{ij-1,,k-1}\right]\right)
\end{aligned} \tag{A37}$$

References

Almatroushi A, Borhan A (2004) Surfactant effect on the buoyancy-driven motion of bubbles and drops in a tube. Ann N Y Acad Sci 1027:330–341

Alves S, Orvalho S, Vasconcelos J (2005) Effect of bubble contamination on rise velocity and mass transfer. Chem Eng Sci 60:1–9

Artoli A (2003) Mesoscopic computational hemodynamics. University Van Amsterdam, The Netherlands, pp 1–142

Ascoli EP, Dandy DS, Leal LG (1990) Buoyancy-driven motion of a deformable drop toward a planar wall at low Reynolds number. J Fluid Mech 213:287–311

Baskurt O, Meisleman H (2003) Blood rheology and hemodynamics. Semin Thromb Hemost 29:435–450

Bedeaux D (1983) The effective shear viscosity for two-phase flow. Phys A Stat Theor Phys 121:345–361

Bel Fdhila R, Duineveld PC (1996) The effect of surfactant on the rise of a spherical bubble at high Reynolds and Peclet numbers. Phys Fluids 8:310–321

Bhaga D, Weber M (1981) Bubbles in viscous liquids: shapes, wakes, and velocities. J Fluid Mech 105:61–85

Boryczko K, Dzwinel W, Yuen D (2003) Dynamical clustering of red blood cells in capillary vessels. J Mol Model 9:16–33

Braasch D (1971) Red cell deformability and capillary blood flow. Physiol Rev 51:679–701

Buick J, Greated A (2000) Gravity in a Lattice Boltzmann model. Phys Rev E 61:5307–5320

Chandran KB, Rittgers SE, Yoganathan AP (2007) Biofluid mechanics: the human circulation, 1st edn. CRC, Abingdon, UK, Ch. II

Chandran K, Yoganathan A, Rittgers S (2006) Biofluid mechanics the human circulation. CRC, Abingdon, UK, pp 116–150

Chauveteau G, Omari A, Tabary R, Renard M, Rose J (2000) Controlling gelation time and microgel size for water shutoff. SPE 59317 SPE/DOE improved oil. doi: 10.2118/59317-MS

Chen H, Chen S, Matthaeus WH (1992) Recovery of the Navier-Stokes equations using a lattice Boltzmann method. Phys Rev A 45:R5339

Cheng J, Xu S, Wen L, Chen J (2005) Steric repulsion between internal aqueous droplets and the external aqueous phase in double emulsion. Langmuir 21:12047–12052

Clift R, Grace J, Weber M (2005) Bubbles drops and particles. Dover, Mineola, NY, pp 1–380

Cox RG (1969) The deformation of a drop in a general time-dependent fluid flow. J Fluid Mech 37:601–623

Cristini V, Hooper R, Macosko C (2002) A numerical and experimental investigation of lamellar blend morphologies. Ind Eng Chem Res 41:6305–6311

Derksen JJ, Eskin D (2010) Potential of microchannel flow for agglomerate breakage. Ind Eng Chem Res 49:10633–10640

H. Farhat et al., *Accelerated Lattice Boltzmann Model for Colloidal Suspensions: Rheology and Interface Morphology*, DOI 10.1007/978-1-4899-7402-0,

D'Ortona U, Salin D, Cieplak M, Rybka R, Banavar J (1995) Two-color nonlinear Boltzmann cellular automata: surface tension and wetting. Phys Rev E 51:3718–3728

Driessen G, Haest C, Heidtmann H, Kamp D, Schmid-Schonbein H (1980) Effect of reduced red cell deformability on flow velocity in capillaries of rat mesentery. Eur J Physiol 388:75–78

Drumright-Clarke MA (2002) Numerical simulations that characterize the effects of surfactant on droplets in shear flow. Virginia Polytechnic Institute and State University Thesis. pp. 1–81

Drumright-Clarke MA, Renardy Y (2004) The effect of insoluble surfactant at dilute concentration on drop breakup under shear with inertia. Phys Fluids 16:14–21

Dupin M, Halliday I, Care C (2003) Multi-component lattice Boltzmann equation for mesoscale blood flow. J Phys A Math Gen 36:8517–8534

Dupin M, Halliday I, Care C (2005) A multi-component lattice Boltzmann scheme: towards the mesoscale simulation of blood flow. Med Eng Phys 28:13–18

Dupin M, Halliday I, Care C, Alboul L, Munn L (2007) Modeling of the flow of dense suspension of deformable particles in three dimensions. Phys Rev E 75:066707-1–066707-17

Eggleton C, Tsai T, Stebe K (2001) Tip streaming from a drop in the presence of surfactants. Phys Rev Lett 87:048302-1–048302-4

Escudier MP, Gouldson IW, Pereira AS, Pinho FT, Poole RJ (2001) On the reproducibility of the rheology of shear-thinning liquids. J Non-Newton Fluid Mech 28:99–124

Farhat H, Celiker F, Singh T, Lee JS (2011) A hybrid lattice Boltzmann model for surfactant-covered droplets. Soft Matter 7:1968–1985

Feigl K, Megias-Alguacil D, Fischer P, Windhab E (2007) Simulation and experiments of droplet deformation and orientation in simple shear flow with surfactants. Chem Eng Sci 62:3242–3258

Filippova O, Hanel D (1998) Grid refinement for lattice-BGK models. J Comput Phys 147:219–228

Gedde M, Yang E, Huestis W (1995) Shape response of human erythrocytes to altered cell PH. Blood 86:1595–1599

Grace J (1973) Shapes and velocities of bubbles rising in infinite liquids. Trans Inst Chem Eng Sci 51:116–120

Griffith R (1962) The effects of surfactants on the terminal velocity of drops and bubbles. Chem Eng Sci 17:1057–1070

Grunau D, Chen S, Eggert K (1993) A lattice Boltzmann model for multiphase fluid flows. Phys Fluids A 5:2557–2561

Gunstensen A, Rothman D, Zaleski S, Zanetti G (1991) Lattice Boltzmann model of immiscible fluids. Phys Rev A 43:4320–4327

Guo Z, Shi B, Wang N (2000) Lattice BGK model for incompressible Navier-Stokes equation. J Comput Phys 165:288–306

Guo Z, Zheng C, Shi B (2002) Discrete lattice Boltzmann effects on the forcing term in the lattice Boltzmann method. Phys Rev E 65:046308-1–046308-6

Gupta A, Kumar R (2008) Lattice Boltzmann simulation to study multiple bubble dynamics. Int J Heat Mass Transf 51:5192–5203

Halliday I, Hollis A, Care C (2005) Improved simulation of drop dynamics in a shear flow at low Reynolds number and capillary number. Phys Rev E 73:1–11

Halliday I, Hollis A, Care C (2007) Lattice Boltzmann algorithm for continuum multicomponent flow. Phys Rev E 76:026708-1–026708-13

Halliday I, Law R, Care C, Hollis A (2006) Improved simulation of drop dynamics in a shear flow at low Reynolds and capillary number. Phys Rev E 73:056708-1–056708-11

Hashim Z, Shtrikman S (1963) A variational approach to the theory of the elastic behaviour of multiphase materials. J Mech Phys Solids 11:127–140

He X, Doolen G (1997) Lattice Boltzmann method on curvilinear coordinates system: flow around a circular cylinder. J Comput Phys 134:306–315

He X, Luo L, Dembo M (1996) Some progress in lattice Boltzmann method. Nonuniform mesh grids. J Comput Phys 129:357–363

Hodges S, Jensen O, Rallison J (2004) Sliding, slipping, and rolling: the sedimentation of viscous drop down a gently incline plane. J Fluid Mech 512:95–131
Hollis A, Halliday I, Law R (2007) Kinematic condition for multicomponent lattice Boltzmann simulation. Phys Rev E 76:026709-1–026709-6
Hong Jeong J, Suggii Y, Minamiyama M, Okamoto K (2006) Measurement of RBC deformation and velocity in capillaries in vivo. Microvasc Res 71:212–217
Hsueh C, Wei W (2009) Analyses of effective viscosity of suspensions with deformable polydispersed spheres. J Phys D Appl Phys 42:075503
Hu T, Lips A (2003) Estimating surfactant surface coverage and decomposing its effect on drop deformation. Phys Rev Lett 91:044501–4
Huang D, Lee TS, Shu C (2007a) Hybrid lattice Boltzmann finite-difference simulation of axisymmetric swirling and rotating flows. Int J Numer Method Fluids 53:1707–1726
Huang H, Thorne D Jr, Schaap M, Sukop M (2007b) Proposed approximation for contact angles in Shan-and-Chen-type multicomponent multiphase lattice Boltzmann models. Phys Rev E 76:066701-1–066701-6
Inamuro T, Suzuki K, Nakamura T, Yoshida M (2005) Acceleration of steady-state lattice Boltzmann simulations on non-uniform mesh using local time step method. J Comput Phys 202:645–663
James AF, Wilson SDR (1978) The film drainage problem in droplet coalescence. J Fluid Mech 87:263–288
Jeon H, Macosko C (2003) Visualization of block copolymer distribution on a sheared drop. Polymer 44:5381–5386
Jones AF, Wilson SDR (1978) The film drainage in droplet coalescence. J Fluid Mech 87:263–288
Kandhai D, Soll W, Chen S, Hoekstra A, Sloot P (2000) Finite-difference lattice-BGK methods on nested grids. Comput Phys Commun 129:100–109
Keller S, Pitcher W III, Huestis W, McConnell H (1998) Red blood cell lipids form immiscible liquids. Phys Rev Lett 81:5019–5022
Kleshchanok D, Lang P (2007) Steric repulsion by adsorbed polymer layers studied with total internal reflection microscopy. Langmuir 23:4332–4339
Kruijt-Stegeman Y, van de Vosse F, Meijer H (2004) Droplet behavior in the presence of insoluble surfactants. Phys Fluids 16:2785–2796
Ladd AJC (1990) Hydrodynamic transport coefficients of random dispersions of hard spheres. J Chem Phys 93:3484–3495
Lai M, Tseng Y, Huang H (2008) An immersed boundary method for interfacial flows with insoluble surfactant. J Comput Phys 227:7279–7293
Latt J (2007) Hydrodynamic limit of the lattice Boltzmann equations. University of Geneva Thesis. pp. 1–124
Legendre D, Magnaudet J (1998) The lift force on spherical bubble in a viscous linear shear flow. J Fluid Mech 368:81–126
Li X, Pozridikis C (1997) The effect of surfactants on drop deformation and on the rheology of dilute emulsions in Stokes flow. J Fluid Mech 341:165–194
Li Y, LeBoeuf E, Basu PK (2005) Least-squares finite-element scheme for the lattice Boltzmann method on an unstructured mesh. Phys Rev E 72:046711-1–046711-11
Lin C, Lai Y (2000) Lattice Boltzmann method on composite grids. Phys Rev E 62:2219–2225
Lipowsky R, Sakmann E (1995) Physical actions in biological adhesion. Handbook of biological physics, 15. Elsevier, Amsterdam, pp 724–753
Lishchuk S, Care C, Halliday I (2003) Lattice Boltzmann algorithm for surface tension with greatly reduced microcurrents. Phys Rev E 67:036701-1–036701-5
Lishchuk S, Halliday I, Care C (2008) Multicomponent lattice Boltzmann method for fluids with a density contrast. Phys Rev E 77:036702-1–036702-8
Lister J, Morrison N, Rallison J (2006) Sedimentation of two-dimensional drop toward rigid horizontal plane. J Fluid Mech 552:345–351

Liu H, Zhou JG, Burrows R (2009) Multi-block lattice Boltzmann simulations of subcritical flow in open channel junctions. Comput Fluids 38:1108–1117

Liu Y, Liu W (2006) Rheology of red blood cell aggregation by computer simulation. J Comput Phys 220:139–154

Lyu S, Jones T, Bates F, Macosko C (2002) Role of block copolymers in suppression of droplet coalescence. Macromolecules 35:7845–7855

Maini D (2007) VOF Based Multiphase lattice Boltzmann method using explicit kinematic boundary conditions at the interface. University of Georgia Thesis. pp. 1–72

Martys M, Chen H (1996) Simulation of multi-component fluids in complex three-dimensional geometries by the lattice Boltzmann method. Phys Rev E 53:743–750

Mason TG, Bibette J, Weitz DA (1996) Yielding and flow of monodisperse emulsions. J Colloid Interface Sci 72:439–448

Milliken WJ, Leal LG (1991) Deformation and breakup of viscoelastic drops in planar extensional flows. J Nonnewton Fluid Mech 40:355–379

Milliken W, Stone H, Leal L (1993) The effect of surfactant on the transient motion of Newtonian drops. Phys Fluids A 5:69–79

Mortazavi S, Tryggvason G (2000) A numerical study of the motion of drops in Poiseuille flow. Part 1. Lateral migration of one drop. J Fluid Mech 411:335–350

Nakano A, Sugii Y, Minamiyama M, Seki J, Niimi H (2005) Velocity profiles of pulsatile blood flow in arterioles with bifurcation and confluence in rat mesnetery measured by particle image velocimetry. JSME Int J 48:444–452

Neu B, Meiselman H (2002) Depletion-mediated red blood cell aggregation in polymer solutions. Biophys J 83:2482–2490

Omari A, Chauveteau G, Tabary R (2003) Gelation of polymer solutions under shear flow. Colloid Surf A Phys Chem Eng Aspects 225:37–48

Ouared R, Chopard B (2005) Lattice Boltzmann simulation of blood flow: non-Newtonian rheology and clotting process. J Stat Phys 121:209–221

Ozawa T, Tanahashi T (2005) CIVA and ANR method for discrete Boltzmann equation. JSME Int J 48:229–234

Pal R (1996) Multiple O/W/O emulsion rheology. Langmuir 12:2220–2225

Pal R (2000) Shear viscosity behavior of emulsions of two immiscible liquids. J Colloid Interface Sci 225:359–366

Pal R (2001) Single-parameter and two-parameter rheological equations of state of nondilute emulsions. Ind Eng Chem Res 40:5666–5674

Peixinho J, Karanjkar PU, Lee JW, Morris JF (2010) Rheology of hydrate forming emulsions. Langmuir 26:11699–11704

Phan-Thien N, Pham D (1997) Differential multiphase models for polydispersed suspensions and particulate solids. J Non-Newton Fluid Mech 72:305–318

Reis T, Philip T (2007) Lattice Boltzmann model for simulating immiscible two-phase flows. J Phys A Math Theor 40:4033–4053

Rorres C, Howard A (1984) Applications of linear algebra, 3rd edn. Wiley, New York

Rostam S, Mahdi M (2009) Solution of a system of two-dimensional linear Fredholm integral equation of the second kind by quadrature methods. Aust J Basic Appl Sci 3:1701–1715

Schafer M, Turek S (1996) Benchmark computations of laminar flow around a cylinder. Notes Numer Fluid Mech 52:547–566

Secomb T, Hsu R, Pries A (1998) A model for red blood cell motion in glycocalyx-lined capillaries. Am J Physiol Heart Circ 274:1016–1022

Secomb T, Hsu R, Pries A (2001) Motion of red blood cell in a capillary with an endothelial surface layer: effect of flow velocity. Am J Physiol Heart Circ 281:629–636

Seifert U (1999) Hydrodynamic lift on bound vesicle. Phys Rev Lett 83:876–879

Shan X, Chen H (1993) Lattice Boltzmann model for simulating flows with multiple phase and components. Phys Rev E 47:1815–1819

Shan X, Chen H (1994) Simulation of non-ideal gases and liquid-gas phase transitions by the lattice Boltzmann. Phys Rev E 49:2941–2948

Shirani E, Jafari S (2007) Application of LBM in simulation of flow in simple micro-geometries and micro porous media. Afr Phys Rev 1:34–42
Shu C, Chew YT, Niu XD (2001) Least-squares-based lattice Boltzmann method: a meshless approach for simulation of flows with complex geometry. Phys Rev E 64:045701-1–045701-4
Sukop MC, Thorne DT Jr (2006) Lattice Boltzmann modeling: an introduction for geoscientists and engineers. Springer, Berlin
Sukumaran S, Siefert U (2001) Influence of shear flow on vesicles near a wall: a numerical study. Phys Rev E 64:011916-1–011916-11
Sun C, Munn L (2006) Influence of erythrocyte aggregation on leukocyte margination in postcapillary expansions: a lattice Boltzmann analysis. Phys A 362:191–196
Sun C, Munn L (2005) Particulate nature of blood determines macroscopic rheology: a 2-D lattice Boltzmann analysis. Biophys J 88:1635–1645
Sundararaj U, Macosko C (1995) Drop breakup and coalescence in polymer blends: the effects of concentration and compatibilization. Macromolecules 28:2647–2657
Takada N, Misawa M, Tomiyama A, Hosokawa S (2001) Simulation of bubble motion under gravity by lattice Boltzmann method. J Nucl Sci Technol 28:330–341
Tannehill J, Anderson D, Pletcher R (1997) Computational fluid mechanics and heat transfer, 2nd edn. Taylor & Francis, Washington, DC
Tasoglu S, Demicri U, Muradoglu M (2008) The effect of soluble surfactant on the transient motion of a buoyancy-driven bubble. Phys Fluids 20:1–15
Tolke J, Freudiger S, Krafczyk M (2006) An adaptive scheme using hierarchical grids for lattice Boltzmann multiphase flow simulations. Comput Fluids 35:820–830
Tolke J, Krafczyk M, Schulz M, Rank E (2002) Lattice Boltzmann simulations of binary fluid flow through porous media. Philos Trans A Math Phys Eng Sci 360:535–545
Tolpekin VA, Duits MHG, van den Ende D, Mellema J (2004) Aggregation and breakup of colloidal particle aggregates in shear flow. Langmuir 20:2614–2627
Tsubota K, Wada S, Yamaguchi T (2006) Simulation study on effects of hematocrit on blood flow properties using particle method. J Biomech Sci Eng 1:159–170
van der Graaf S (2006) Membrane emulsification: droplet formation and effects of interfacial tension. Wageningen University, Amsterdam, Thesis. pp. 1–159
van der Sman R (2004) Diffusion on unstructured triangular grids using lattice Boltzmann. Future Generat Comput Syst 20:965–971
van der Sman R (2003) Lattice Botlzmann scheme for diffusion on triangular grids, Computational science-ICCS 2003. Springer, Berlin
van der Snam R, van der Graaf S (2006) Diffuse interface model of surfactant adsorption into flat and droplet interfaces. Rheol Acta 46:3–11
Velankar S, Zhou H, Kyoung Jeon H, Macosko C (2004) CFD evaluation of drop retraction methods for the measurement of interfacial tension of surfactant-laden drops. J Colloid Interface Sci 272:172–185
Vlahovska P (2003) Dynamics of surfactant-covered drop and the non-Newtonian rheology of emulsions. Yale University Thesis. pp. 1–152
Vlahovska P, Loewenberg M, Blawzdziewicz J (2005) Deformation of surfactant covered drop in a linear flow. Phys Fluids 17:103103-1–103103-18
Williams A, Janssen J, Prins A (1997) Behaviour of droplets in simple shear flow in the presence of protein emulsifier. Colloidal Surf A 125:189–200
Yiantsio SG, Davis RH (1990) On the buoyancy-driven motion of a drop towards a rigid surface or a deformable interface. J Fluid Mech 217:547
Yu D, Girimaji S (2006) Multi- block lattice Boltzmann method: extension to 3D and validation in turbulence. Phys A 63:118–124
Yu D, Mei R, Shyy W (2002) A multi-block lattice Boltzmann method for viscous fluid flows. Int J Numer Meth Fluid 39:99–120
Zou Q, He X (1997) On pressure and velocity boundary conditions for the lattice Boltzmann BGK model. Phys Fluids 9:1591–1598

Index

H. Farhat et al., *Accelerated Lattice Boltzmann Model for Colloidal Suspensions: Rheology and Interface Morphology*, DOI 10.1007/978-1-4899-7402-0,